THE ENTHUSIAST'S GUIDE TO

BRITISH POSTWAR CLASSIC CARS

JONATHAN WOOD

THE ENTHUSIAST'S GUIDE TO

BRITISH POSTWAR CLASSIC CARS

JONATHAN WOOD

OSPREY

Published in the winter of 1980 by
Osprey Publishing Limited
12–14 Long Acre, London WC2E 9LP
Member of the George Philip Group

British Library Cataloguing in Publication Data
Wood, Jonathan
The enthusiast's guide to classic cars
1. Automobiles – Great Britain – History
I. Title
629.22′22′0941 TL57

ISBN 0-85045-377-1

Editor Tim Parker
Filmset and printed in England by
BAS Printers Limited,
Over Wallop, Hampshire

Front cover: left to right, 1960 long wheelbase Rolls-Royce Silver Cloud II (Frank Dale and Stepsons Ltd.); 1949 Morris Minor, Series MM (Derek Porter); 1965 Jaguar E-type, Series I (Will Athawes)

CONTENTS

INTRODUCTION

This book on classic cars will, I trust, appeal to the enthusiast, the prospective purchaser and the historian. The vehicles included are all of British manufacture and the models in question were introduced between 1945 and 1967.

My starting point of 1945 presented no problems as it marked the end of the Second World War and signalled the re-starting of private car production in Britain. The difficulty lay in deciding on an appropriate finishing point! I eventually opted for 1967 for the simple reason that the following year witnessed the creation of the British Leyland Motor Corporation; there can be few who would argue that our industry's affairs have been quite the same since then. Also, in January 1968, more rigorous American safety regulations came into force, a move that did, and was destined to, profoundly influence the shape of many of our cars. The demise of the 'Big' Healey, in the face of these new transatlantic requirements, perhaps symbolizes the ending of that era.

Now to the format of the book itself. You will find it divided into two sections. In the first part I have endeavoured to trace the evolution of the British motor industry to 1968 and to dwell on the personalities who, I believe, shaped our postwar destiny. Inevitably no chronicle of the '40s, '50s and '60s would be possible without considering the formative pre-war years, particularly the 1930s, a decade that witnessed the emergence of the 'Big Six' motor manufacturers.

In the second section, which makes up the bulk of the book, I have chosen what I hope is a fair cross-section of those postwar British cars that have achieved collectable status. But this is *not* intended to be a definitive list of classic cars. Each marque entry is sub-divided into three parts. Firstly, I have set down the particular company's history, with the accent on the postwar years. Then, secondly, I have selected an appropriate model, or models, and detailed the production span and noted changes in specification. Thirdly comes what I have called *check points*, in other words, the 'Bad News'! This is intended as a guide for those contemplating the model in question and who want to know the snags. . . . For this section I have followed an approach adopted for the *Classic Choice* articles that have proved such a popular feature of *Thoroughbred and Classic Cars*. At this point my thanks must go to the following for extra information: Peter Caunt, G. A. Horrox, Armstrong Siddeley Owners' Club (Armstrong Siddeley Sapphire); Richard Williams, Richard Stewart Williams Ltd., (Aston Martin DB 4/5/6); E. J. Wooddin, John Juleff, Bristol Owners' Club (Bristol 400–406); Peter Wallis, Rex Woodham-Harrison, Jensen Owners' Club (Jensen 541); the late G. F.

Simonds, Sunbeam–Talbot Alpine Register (Sunbeam–Talbot and Alpine); John Hanna, TR Register (TR4A).

In addition, I am most indebted to a host of one-make historians and fellow scribes, who have assisted with the individual marque histories. My grateful thanks to: Warren Allport, Michael Bowler, Anders Clausager, Colin Cowey, Kenneth Day, John Hubbuck, Brian Joscelyne, John McLellan, Colin Musgrove, Doug Nye, Brian Palmer, John Pellowe, Graham Robson, Paul Skilleter, Brian Smith and David Burgess Wise.

I should point out that I have, in the main, referred to cars by their model year (e.g. announced at 1955 Motor Show but therefore 1956 model).

My thanks to those who loaned cars for the front cover of this book. They are Will Athawes (Jaguar E type), John Gordon of Frank Dale and Stepsons (Rolls–Royce Silver Cloud) and Derek Porter (Morris Minor).

Also included are production details and performance/consumption figures. For the latter I have drawn on contemporary *Autocar* and *Motor* road tests and my grateful thanks for permission to reproduce them must go to the respective editors of both publications. I should, perhaps, qualify the *top speed* figures, as this refers to mean speed, obtained by four opposite timed runs.

In addition, there are the names of just some of the companies who specialize in providing spares and restoration services for the cars' featured, though I must stress that these are listings only, not recommendations.

As far as the book's illustrations are concerned, contemporary photographs have been used throughout which I trust will also prove to be a useful restoration aid, where originality should be the watchword.

Finally my thanks must go to my colleague, Lionel Burrell, for not only being responsible for the concept of *Thoroughbred and Classic Cars*, but also for suggesting this book's format to me. And to Tim Parker, Osprey's managing editor, goes my appreciation for his encouragement.

You may run a mass-produced saloon or a handcrafted two seater but whatever your preferences, I hope that there will be something of interest within these pages.

Jonathan Wood

Farnham, Surrey, May 1980

PART 1

BRITAIN AND THE MOTOR CAR

The religion of the Company lies in the successful product. . . . Terry Beckett is 'Mr Cortina' and became chairman of Ford of Britain.
Edouard Seidler, LET'S CALL IT FIESTA.*

'What's this bloody contraption?' I drove him round Longbridge flat out. He shouted 'stop!' — got out — said 'make the bloody thing' and walked away.
Sir Alec Issigonis's recollection of Sir Leonard Lord sanctioning the production of the Mini.
THOROUGHBRED AND CLASSIC CARS, *September 1979.*

Before the Second World War

The 22 years following the end of the Second World War have proved to be some of the most turbulent in the history of the British motor industry. For, if the 1950s was a time of expansion and self-confidence, then the next decade was one of re-trenchment and merger. In 1945 our major car manufacturers could be dubbed the 'Big Six'. By 1968 there were Four.

These were years that saw Britain relegated to fourth position in the world car manufacturing stakes, having been second in the league since the early '30s. The era culminated with the take-over of the country's largest car manufacturing group by Leyland Motors — a move that led to eventual nationalization.

Yet on a more positive front these were years of rapid expansion for the Ford Motor Company, an organization that skilfully caught the public mood and profitably satisfied it. This was a time that saw the consolidation of the Jaguar marque with Sir William Lyons emerging as the outstanding figure of the postwar years. And the genius of Alec Issigonis created the Mini, the most revolutionary car in the history of the British motor industry.

The postwar era saw the virtual end of the traditional craft of coachbuilding and the arrival of the glass-fibre body. Then came dominance in the field of Grand Prix racing and the birth of the Lotus marque with its creator, Colin Chapman, appearing as a major force in the land. But this was a time that witnessed the gradual demise of the open car.

To try and find reasons for the successes, failures and contradictions of the postwar years we must return to the 1930s, as it was during this decade that the Big Six motor manufacturers emerged, setting the stage for the competitive rigours of the late '50s and '60s.

*Patrick Stephens, 1976

Morris's first car was the two seater Oxford, introduced in 1913. It was a skilful assemblage of bought-out parts. Chassis was by Rubery Owen, wheels by Sankey and axles from E. G. Wrigley. The engine

The expansionist 1930s

If the pattern of the 1920s had been repeated during the following decade the Morris and Austin companies would have been in an unassailable position by the outbreak of the Second World War in 1939. Ten years previously Morris, Austin and Singer accounted for no less than 75 per cent of the market. Yet by 1938 the two leaders' market share had been seriously eroded and Singer had largely faded from the picture. Morris was only just ahead in production terms, closely pursued by Austin, Ford, Vauxhall, Rootes and Standard, accounting for 90 per cent of British output. The remaining 10 per cent was made up of the specialist companies headed by Rover, which had been steered to commercial viability by Spencer Wilks, and followed by Singer, BSA (including Daimler and Lanchester), S.S. which had been created by William Lyons in the very teeth of the 1931

depression, Armstrong Siddeley, Jowett and with Rolls–Royce/Bentley and Alvis completing the picture. Sports cars were produced, by comparison, in penny numbers. MG was the most productive company, each of these inexpensive but rakish two-seater models usually selling between 2000 and 3000 examples.

The decade opened to a world dominated by depression with initially the manufacturers of low-production, high-quality cars suffering from an erosion of their markets. Bentley were taken over in 1931 by Rolls–Royce, who were safely underpinned by their growing aero-engine sales, and BSA absorbed Lanchester while Lagonda staggered on until it too collapsed in 1935 and was then reconstructed in an improved financial climate. The prestigious Sunbeam and Talbot companies, embroiled in the demise of the unwieldy Sunbeam–Talbot–Darracq coalition, were drawn into the Rootes brothers' expanding empire in 1935. Riley succumbed to liquidation in 1938 and became part of Morris Motors, while Triumph went out of business the following year.

But these failures should be seen against a background of overall expansion with British car production doubling between 1928 and 1938 and the country developing as Europe's largest motor manufacturer by 1932, being then second only to the United States of America in vehicle output. The '30s was the era of the appearance of the small car and, if the 11.9 HP 'Bullnose' Morris Cowley typified the previous decade, then perhaps the Ford Eight speaks for these later years.

There can be little doubt that the dominating personalities of these inter-war years were William Richard Morris (1877–1963) and Herbert Austin (1866–1941), who began selling cars under their own names in 1913 and 1906 respectively. Both had agricultural backgrounds, were largely self-educated in technical matters, but employed their powers of determination, drive and intuition to the full. It is necessary to examine the weaknesses, as well as the strengths, of their careers because the courses that they took have shaped the fabric of the British motor industry to this very day.

and gearbox came from White and Poppe, the body Raworth and radiator by Coventry Motor Fittings. The Standard (shown above) and more expensive de Luxe models remained available until 1916

William Morris

William Morris was born in Comer Gardens, Worcester. Both parents were of Oxfordshire farming stock, though Morris senior had pursued a varied career which had included crossing the Atlantic and driving a Royal Mail coach in Canada! Frederick Morris was a draper in Worcester, but when William was 3 years of age the family moved back to their native Oxfordshire when Morris senior became bailiff to his blind father-in-law, who farmed at Headington Quarry, near Oxford. Morris attended St. James Church of England School at Cowley and he left at 15 years of age, faced with getting a job because of his father's ill health. This was the only education he ever received and, although he signed on for evening classes in engineering, he only attended twice. As Morris displayed a natural interest in things mechanical, he went to work in an Oxford bicycle shop in 1893 but soon left when he failed to secure an increase in salary. He therefore set up his own business repairing bicycles in his father's garden shed. It wasn't long before he began building his own machines and his first customer was the local vicar, whose lanky proportions demanded a 28-inch frame! Bicycle construction was a fairly straightforward business as he purchased and then assembled components from specialist manufacturers. Consequently his financial outlay was comparatively small as was the risk involved.

The business, such as it was, certainly expanded. In 1901 Morris even produced a motorcycle and the following year he entered a short-lived partnership with Joseph Cooper, a local bicycle dealer. Then, in 1903, he was persuaded by a wealthy undergraduate to become works' manager of the grandly-titled Oxford Automobile and Cycle Agency, the intention being to market motor cars and manufacture bicycles and motorcycles. However, the business went bankrupt within a year and Morris was on his own again as a motor agent and cycle maker. These were valuable years as he could study car design at first hand and gauge public response to individual models. And above all he acquired a first-hand knowledge of the individual firms supplying the rapidly-expanding British motor industry. By 1910 he was investigating the sources of supply of motor car components with a view to producing his own car. Morris intended to adopt the same principles that he had employed in producing bicycles. By relying on specialists to provide parts for his car, rather than manufacture them himself, he could keep his costs to a minimum and accordingly benefit from the latest engineering developments. But, above all, he still lacked financial resources. Then, in 1911, the young Earl of Macclesfield agreed to invest £4000 in Morris's new business, and W.R.M. Motors Ltd, was founded the following year. Morris was on his way. He went to the Coventry firm of White and Poppe for his engine/gearbox unit, E. G. Wrigley of Birmingham provided axles, Sankey metal artillery wheels, while Raworth, a local Oxford coachbuilder, supplied bodies. Morris then needed somewhere to assemble his cars, so took over premises at Temple Cowley. Although the building in question had been a military training college, previously it had been occupied by Hurst's Grammar School, which his father had once attended.

White and Poppe hadn't completed their parts by the time of the 1912 Motor Show so Morris attended, armed with a set of blueprints, and London motor agents Stewart and Ardern ordered 400 cars on the strength of these drawings. The car, appropriately called the Morris Oxford, appeared in the spring of 1913 and was competitively priced at £175. But there was a shortcoming in that it could only be built in 2–seater form. If a 4–seater was to be produced then a larger engine than the 1018cc White and Poppe 4–cylinder would be needed. What Morris wanted was to produce a cheap family 4–seater which was just the type of car being successfully mass-produced in America by Henry Ford. As Ford's achievement had such a profound effect on William Morris, and later Herbert Austin and Europe's principal motor manufacturers, we must briefly break off our narrative to see how Henry, another farmer's son, succeeded in putting the world on wheels.

Henry Ford (1863–1947) grew up on a farm in Dearborn, Michigan and

Above, **Henry Ford and his wife Clara pictured on board the White Star liner** ***Majestic*** **when it docked at Southampton on Good Friday, 1928.** ***Right***, **the Model T Ford's engine. Pace setting features are the detachable cylinder head (secured by bolts), one piece cylinder block combined with crankcase and gearbox in unit with the engine. Note the ingenious two speed epicyclic gearbox, flywheel magneto and splash lubrication**

like Morris displayed an early flair for mechanics. He built his first car in 1896 and, with initial support from Alexander Malcomson, a Detroit coal merchant, established the Ford Motor Company in 1903. His Model A appeared that year and a succession of models followed until he hit the automotive jackpot with his Model T of 1908. It was a spidery, yet deceptively tough car, powered by a rugged 2.8–litre, 4–cylinder engine. This revolutionary power unit employed a block and crankcase cast in one piece (rather than the two items being produced separately and then bolted together) and at Ford's insistence, although it flew in the face of convention, the cylinder head was detachable. The two speed epicyclic gearbox was in unit with the engine, instead of following established trends. This single engine/gearbox package thus accelerated the manufacturing process. Extensive use was made of vanadium steel, a strong but light alloy which greatly contributed to the Model T's legendary qualities. When it first went on sale in 1908, the car sold for $850. The following year Ford took the momentous step of dropping all his other models and producing just the Model T. The next step was to try and lower the price, so Henry and his lieutenants set about changing the production process. They were already benefiting from the production of large quantities of interchangeable parts, already pioneered by Oldsmobile in 1901–1904, an inheritance of America's progressive and buoyant machine tool industry. Then, in 1913, Ford adopted a conveyor-belt system for building up the Model T's flywheel magneto

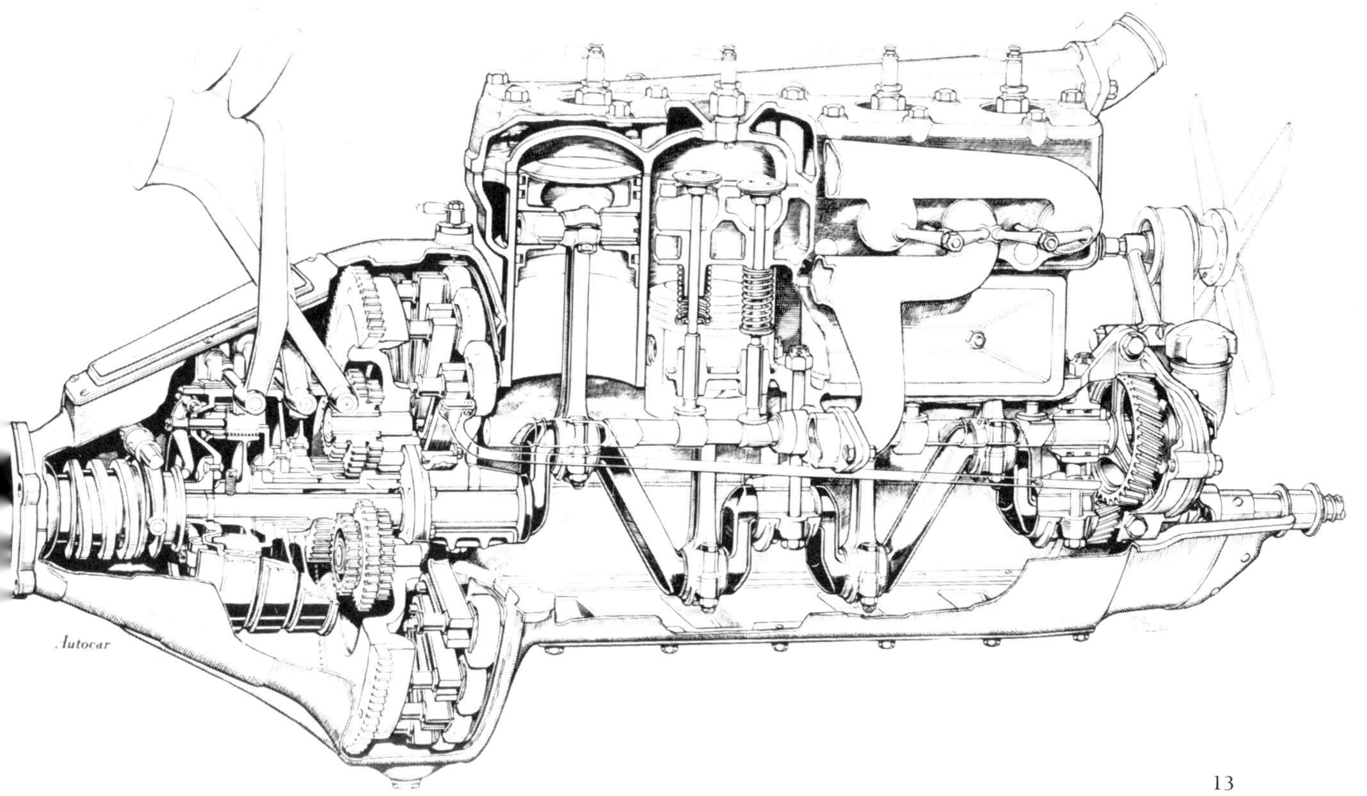

The Ford Model T's moving track assembly line at Highland Park, Detroit. This picture from *Ford Methods and the Ford Shops* shows chassis leaving the rails on which they had previously slid to run on their own wheels. By this time complete chassis assembly had been reduced to an hour and a half. The date is 1914

assembly. This proved highly successful and later that year a crude experiment was introduced whereby a chassis was pulled through the factory on a rope and windlass, components being added as it progressed. The results were astounding. Previously the chassis had remained static and assembly had taken 12 hours and 28 minutes. But with it moving, manufacturing time could be cut by half. Ford had the right car for the right market; he could now produce it in the right way. Car production by moving track was born, output soared and the Model T's price tumbled. It was down to $490 in 1914 and had fallen again by 1916 to $360, when over half a million Model Ts were produced for the first time in one year. Ford had perpetuated the well-established American practice of taking a luxury commodity, streamlining its manufacture and making it available to the man in the street, who then regarded it as a necessity.

With such revolutionary developments taking place in America, it is hardly surprising that at the end of 1913 William Morris crossed the Atlantic, like his father had done, though his objectives were defined and clear cut. He felt that his projected 4–seater Morris would be uncompetitive against American imports and wanted to find out how their lower price was achieved. With him went Hans Lanstad, White and Poppe's Norwegian

The Bullnose Morris, Britain's top seller of the 'twenties. This car was successfully driven non-stop from Lands End to John O'Groats by its owner, Mr. Towser. The proud driver is seen with William Morris (left) and Miles Thomas (right), who had recently joined Morris Motors as publicity adviser. The date is August 28 1924 and the location the yard behind the original Morris factory (now occupied by the Nuffield Press)

chief draughtsman. They sought out Detroit, already established as the home of the American motor industry and the Model T Ford, and visited the Continental Motor Manufacturing Company. Morris told them what he wanted and they provided him with drawings of an engine which they thought would be suitable for his needs. Continental's quotation was $85 per engine (£17 9s 3d) and, even with shipping costs, was appreciably cheaper than the £50 White and Poppe had quoted for an engine and gearbox unit for the projected 4–seater Morris. Morris and Landstad returned to England, complete with Continental's blueprints, but there was no way that Peter Poppe's firm could approach Continental's price. Landstad felt that the only answer was to take 6 months leave of absence, return to America, get a job there and find out what made the American System of Manufacture, in other words, mass-production, tick.

Therefore, in August 1914, just 10 days after the outbreak of the First World War, Morris and Landstad again set sail for the New World aboard the *Mauretania*. But this was no leisurely jaunt because Morris had asked the Norwegian to bring his drawing board and, on their first day at sea, (a Sunday!) he went down to Landstad's cabin and together they started work on the design of the vehicle that was destined to be the best-selling British car of the 1920s: the Morris Cowley.

They again journeyed to Continental at Detroit, Landstad was taken on the payroll and Morris ordered sufficient parts for 3000 cars. The power unit was Continental's U-type 1495cc Red Seal unit; like the Model T Ford it boasted a detachable cylinder head, while the oil level dip-stick also had yet to become fashionable in Britain. Gearboxes came from the Detroit Gear and Machine Company and axles and steering boxes were also ordered. Morris then returned to England, and Landstad followed at the end of the year, leaving White and Poppe to join Morris at Cowley. The Norwegian was, by then, well versed in the business of mass-producing components from interchangeable parts.

Components began arriving at Cowley from America at the end of 1914 and, therefore, it was possible for Morris to announce his Continental Cowley, as the new model was called, in April 1915. Not only did the engine reflect its transatlantic origins, but the ball-mounted gear lever was centrally placed at a time when practically every British-made car boasted a right-mounted one. Initially announced as a 2–seater, the new Morris Cowley was impressively-priced at 158 guineas, which was cheaper than the smaller Morris Oxford. Production didn't get under way for some months, but unfortunately for Morris he was overtaken by international events because, in September 1915, Chancellor of the Exchequer Reginald McKenna announced a 33.3 per cent duty on imported cars and parts to allow shipping space to be reserved for more vital commodities. Clocks and watches also suffered. Prior to this, no duty was payable on imported goods, this policy of Free Trade being one of the tenets of *laissez-faire* governmental policies of the nineteenth century. These 'McKenna Duties', as they were called, remained in force at this level right up until 1956, when they were reduced to 30 per cent, though the 1924 Labour government briefly dispensed with them. As a result, the British motor industry was able to operate behind a tariff wall as foreign cars were forced to pay this import premium. It was an incentive, however, for foreign manufacturers to establish factories in Britain, these products then being the right side of the tariff.

The imposition of these Duties caused Morris's prices to spiral and, in March of the following year, the government prohibited such imports altogether, by which time sufficient parts for around 1500 cars had reached Cowley, so a trickle of vehicles was assembled up until 1920. But car production was playing second fiddle to the war work that the company had taken on. Hand grenades, bomb cases and above all, mine sinkers were mass-produced from interchangeable parts, employing principles learnt in Detroit. Therefore, when the First World War came to an end in 1918, W. R. M.

Ford's world car. A Manchester built Model T, complete with canine passenger. This 1913 car has a body by Scott Brothers of Manchester, a concern later absorbed by the Ford company

DJ-249

Motors (the business became Morris Motors the following year) were in a potentially strong position as far as the postwar car market was concerned. They had a modern design and an engine and gearbox which lent itself to mass production, and the public were crying out for cars. The only difficulty was that Continental had informed Morris that they no longer intended to produce their U-type engine as it was too small for the American market. Morris therefore acquired the manufacturing rights and looked around for someone to make the engine for him in England. White and Poppe didn't want the job and eventually he found that Hotchkiss et Cie, who had established a factory in Coventry during the war when they feared that their own French plant might be overrun, were looking for work. They accepted the job and, above all, didn't require a deposit. The Continental engine was, therefore, copied with a few minor modifications, one of them being the introduction of a cork-faced clutch running in engine oil. This feature was suggested by Morris himself, one of the few contributions he made to the detail design of his own cars. Thus, he was able to resume his pre-war practice of producing his cars from ready-made components and in 1919 the new Cowleys and smarter Oxfords began leaving the Cowley works.

But he was up against stiff opposition from the firm that had been the inspiration for many of his ideas. The Ford Motor Company had established an assembly plant in Britain in 1911 and 2 years later it was Europe's largest producer of motor vehicles. In America output continued to spiral, uninterrupted by war, so that by 1920 it was estimated that every other car in the world was a Model T. Ford made no secret of how he achieved his success and his Highland Park factory was open to visitors from all parts of the globe. In 1915 the *Engineering Magazine* published *Fords Methods and the Ford Shops*, a highly detailed description of the Model T's manufacturing process. Here was a blueprint for any motor manufacturer with sufficient resources to imitate. In France, Renault and the new Citroën company followed in the Model T's wheel tracks as did Fiat in Italy and Morris, Austin and, briefly, Bean, in England. Ford's lead seemed unassailable and in Britain, for instance, Model Ts accounted in 1919 for no less than 41 per cent of new car registrations.

Bean was probably the first British manufacturer to take up the Ford concept of the moving track assembly line. This 1920 photograph shows the 11.9 model in production at their factory at Tipton, Staffordshire. The car was based on the pre-war Perry, seen in the background

During the following year British manufacturers got into their stride with the first flush of peace producing a short-lived boom, which collapsed at the end of 1920. The 12 months that followed were what *The Economist* described as 'one of the worst years of depression since the industrial revolution'. Unemployment soared and Morris was hard hit, like his contemporaries, and even spoke of emigrating to Australia. By March, his factory was full of unsold cars and he then took what Lytton Jarman and Robin Barraclough have called 'the most momentous decision ever made in the British motor industry'.* The price of the 4–seater Cowley was slashed by £100 to £425 and the 2–seater went down from £465 to £375. Then, on the eve of that year's Motor Show, Morris acted again. The 4–seater was reduced to £341 and the smaller car to £299. Sales soared and output for 1922 was more than double the previous year and 'Bullnose' Morrises were seen in

**The Bullnose Morris,* Macdonald, 1965

increasing numbers on Britain's roads.

The other factor that worked in Morris's favour was the Finance Act that came into force in January 1921. It is better remembered as the 'horsepower tax' and its imposition had a disastrous effect on the sales of the Model T Ford, which had always enjoyed an enormous price advantage over its contemporaries. In 1919, for instance, a Model T sold for £220, but a Morris Cowley cost £330. But from 1921 cars were taxed at a rate of £1 per RAC horsepower. With its larger big-bored engine, this meant that the American car carried a 22.5HP RAC rating and cost its owner £23 a year to tax. By contrast, the 11.9HP Morris was taxed at a mere £12 *and* it used less petrol.

To see how this state of affairs came about we must briefly return to 1906. It was in that year that the Royal Automobile Club set down their RAC rating formula. This was expressed as $D^2N/2.5$. D represented the cylinder bore, N the number of cylinders, while 2.5 was arbitrarily arrived at 'and will be found reasonably and sufficiently accurate for comparative purposes'. For this was the thinking behind the formula 'so that the public may arrive at the approximate size of engines *in comparison* [my italics] with others'. But the club was quick to point out that the RAC rating is 'Not to be considered as an accurate or scientific calculation of actual horse power.' However, when the

Chancellor of the Exchequer David Lloyd George delivered his 1909 Budget, he announced that the annual car tax was to be calculated on the RAC formula, thus ensuring that the upkeep of the roads was self-financing. Previously vehicles had been mainly taxed on weight. The new sliding scale, along with a loading on petrol, duly became law, an instance being that a 12HP car cost 4 guineas a year to tax while a 60HP monster was charged £21.

This state of affairs remained in force, as we have seen, until 1921, when big-bored American cars in Britain, and particularly the best-selling Model T, were badly hit because the formula only took account of the vehicle's bore and number of cylinders, the stroke not being considered. This legal strait-jacket had a far-reaching effect on engine design because, in an effort to keep their RAC rating as low as possible, and therefore their potential customers' Road Fund Licence, manufacturers opted for small-bore, long-stroke engines. This, then, was why a vehicle's 'horsepower' was stressed when a car was marketed: Austin Seven, Singer Nine, Morris Eight and so on.

With Model T sales tumbling, Morris forged ahead and in 1925 he sold 54,151 cars on the British market, which represented a staggering 41 per cent of total car production. The famous 'Bullnose' finally ceased in the following year to be replaced by the plainer 'Flatnose' radiator model and, for a time, overall expansion was maintained, reaching a peak for the decade in 1929, when 63,522 cars were produced.

As we have seen, Morris's method of building cars was to assemble ready-made components purchased from specialist suppliers. But in 1922, when planning for future expansions, he realized that his growth might be limited by the inability of these firms to keep pace with his demand. Therefore, in 1923, he purchased the Oxford-based Osberton Radiators, the Coventry firm of Hollick and Pratt, who built his bodies, and Hotchkiss et Cie's Coventry works, which had been producing his engines since 1919. These were renamed Radiators Branch, Morris Bodies and Morris Engines respectively. In 1924 he absorbed axle supplier E. G. Wrigley of Birmingham, using their works to expand into commercial vehicle production. Then, in 1926, he bought the SU Carburettor Company. In this manufacturing philosophy Morris was at odds with his great contemporary Herbert Austin, who made a point of making as much as possible at his great works at Longbridge on the southern outskirts of Birmingham.

Herbert Austin, at the age of 40, pictured during the First World War

Herbert Austin

Herbert Austin, a farmer's son, was born at Grange Farm, Little Missenden, Buckinghamshire, in 1866. But when he was a youngster the family moved to Yorkshire, where his father, like Morris's, became a farm bailiff, though in Giles Austin's case the job was on Earl Fitzwilliam's Wentworth estate. Austin attended Rotherham Grammar School and soon displayed an early talent for drawing. 'The long winter evenings would have been quite awful unless I had been able to indulge my hobby', he later remembered. He

therefore decided to pursue an architectural career and after leaving school spent 2 years training at Brampton Commercial College. But his strong natural instinct for mechanical matters made him realize that architecture held no appeal for him. So his parents secured a provisional acceptance for their son at the Great Northern Railway works, but he was sidetracked from this objective by a visit from his mother's brother, who had emigrated to Australia and was in England on a short visit. His uncle convinced him of the prospects that existed for young men in the Antipodes, so Herbert returned to Australia with him. He was 17 years old and served an apprenticeship with Messrs Richard Parks and Co., of Melbourne, where his uncle was manager, though 2 years later he moved on to a company that imported Crossley gas engines and printing presses. A two-year stint with this firm was followed by another move to Langland's Foundry Company. Austin later admitted to having 'mixed memories' of the place but readily conceded, 'it was here that I achieved a thorough training as a mechanic'.

It was while at Langland's that he met up with another immigrant, Frederick York Wolseley from Dublin, whose brother was Field-Marshal Sir Garnet Wolseley. Frederick had been managing a sheep station in New South Wales and, having first-hand experience of the business, was convinced that there would be a ready sale for mechanical sheep-shearing equipment, to replace the usual manual method. In 1887 the Wolseley Sheep Shearing Company was established in Sydney and Frederick cast about for sub-contractors to produce components, which is how he came to meet Herbert Austin. Eventually, in 1889, the Australian end of Wolseley's company was wound up and transferred to London. Austin was later offered the post of general manager and he and his wife (whom he had married in 1887) returned to England in the winter of 1893.

But there were still assembly problems with defective parts flooding back to Britain and Austin realized that the only satisfactory solution was for Wolseley to manufacture the components themselves; in 1895 he initiated the purchase of a Birmingham factory for this very purpose. Sheep-shearing equipment was produced there along with machine tools and bicycle parts and then, probably in 1895, Austin tried his hand at car design. Therefore, along with Frederick Lanchester, he can be seen as a British pioneer, though whereas Lanchester's car was the result of a truly formidable and trained intellect, Austin's more empirical approach relied more on copying an existing design. Herbert's car was based, in outline, on the French Léon Bollée; the flat twin 2HP engine was certainly in advance of that used on the French machine, though the final drive arrangement was a virtual carbon copy. This vehicle was not developed, probably because of the risk of patent infringement, the recently-formed British Motor Syndicate having snapped up the Bollée patents. Austin's next attempt in 1897 was a more conventional three-wheeler, again with a horizontally-mounted engine, but production cars were based on a 3.5HP 4-wheeler, in which he successfully competed in the 1000 Mile Trial of 1900. An approach from the mighty engineering combine of Vickers, Son and Maxim resulted in their acquiring the machine tool and motor car side of the Wolseley company and this new concern was duly registered in 1901, with Herbert Austin as general manager. He remained with the company until the summer of 1905 when, following a disagreement, he walked out and was then sacked. Ostensibly this was for continuing to design cars with old-fashioned horizontally-mounted engines, but when he later manufactured vehicles under his own name, he employed conventional vertically-mounted power units.

In mid-1905, Austin, with three companions, set out in a little 7½HP Wolseley to look at a disused printing works known as White and Pikes about 7 miles south of Birmingham and adjacent to a branch of the Midland Railway. So it was that Longbridge was chosen as the home of the Austin Motor Company. Herbert attended the 1905 Crystal Palace Motor Exhibition with only a set of drawings of his proposed car, just as William Morris was to do 7 years later. Production started in 1906 and the concern prospered to the extent that, in 1914, it became a public company. These Longbridge products of the pre-First World War era were T-head fours and sixes and indelibly stamped with the hallmark that was to typify Austin products in the '20s and '30s. They were certainly reliable but rather old-fashioned in concept. But the bodywork was well conceived and executed and was built on the premises, instead of being sub-contracted to a specialist coachbuilder.

During the First World War the Longbridge works expanded dramatically, a workforce of 2000 in 1914 swelled to a peak of 22,000 in 1916, munitions, vehicle and aeroplane contracts being undertaken for the war effort. During the conflict Austin's manufacturing policy experienced a marked change of emphasis. Prior to the war he had produced a variety of models, his best seller being a sturdy 20HP car. Inspired, no doubt, by Henry Ford's phenomenally successful policy, in 1919 he scrapped the rest of his range and concentrated on just one car, a massive 20HP product of distinctly American appearance, in the spirit of a Hudson 16 he had run during the war. As on the Model T Ford, the engine and gearbox were mounted in unit and like the Morris the gear lever was centrally positioned, though the sliding gate in which it moved was a concession to convention.

This model — the Twenty — crept into production in 1919/20, but the following year witnessed the same sort of problems Morris was experiencing at Cowley. The difference was that the 11.9HP 'Bullnose' was well tailored to the times and Austin's lumbering 20HP wasn't. Finances suffered to an extent that a receiver was appointed in April 1921 and, as a result of the financial mauling the company received, it didn't start paying dividends again until 1929. Austin fought back and a scaled-down version of the overstrong Twenty was rushed into production in 1921. This was the famous Twelve, destined to be the backbone of the Austin range throughout the 1920s. Although the financial climate eased in 1922 and the receiver was discharged, Ernest L. Payton became secretarial director, deputy chairman in 1928 and chairman and deputy managing director on Austin's death in 1941. The other important appointment of the same year was that of Carl Engelbach, who became works director. This triumvirate of Austin, Payton

The famous Austin Seven, a true large car in miniature. This example (registered OK 9633) was road tested by *The Autocar* in 1923. Although originally of 696cc, in March 1923 the Seven's capacity was increased to 747cc and there it remained until the model ceased production in 1939

and Engelbach was largely responsible for steering the company's fortunes through the '20s and '30s.

Austin Seven

But the car that was an even greater success than the faithful Twelve was a really small one, designed at Austin's home Lickey Grange, so as to speed its creation, over the winter of 1921–1922. What emerged was the Austin Seven that went into production in the latter year, and built up until 1939, by which time over 290,000 had been made. It was to be the best-selling British car of the inter-war years. The Seven's success lay in the fact that it was a large car scaled down, boasting a diminutive 748cc 4–cylinder engine which was the smallest-capacity British built four when it appeared. It followed the French trend for small fours established prior to the First World War and was directly inspired by the 668cc Peugeot Quadrilette. The engine, which was at the core of the model's appeal, was conceived by an 18-year-old draughtsman, Stanley Edge, seconded by his chief from the Longbridge drawing office to work on the project. For the Seven was a great improvement over the noisy and rough-running 2–cylinder cars more usually associated with that sector of the market. But if Austin had had his way the Seven could have ended up as a two-lunger because his original idea for the design was to copy, as was his way, an existing model. He had in mind the air-cooled 2–cylinder Rover Eight. Fortunately Edge's direct approach, coupled by technical knowledge, overcame Austin's essentially empirical outlook and Stanley got his way. The Seven was an outstanding success, paving the way to the small-capacity fours of the 1930s.

With a range of well-built and reliable models, Austin was to continually challenge the cars from Cowley in the 1920s and the timely introduction of a 10Hp car followed in 1932. Relations between Austin and Morris were chilly, to say the least. In 1924 when the first Labour government lifted the McKenna Duties, it was feared that the market would again be flooded by American imports. Dudley Docker of Wolseley therefore proposed that his company, along with Morris and Austin, should amalgamate to face the challenge. Although Docker and Austin were in favour of the idea, Morris could see little advantage, even though it was proposed that he head the alliance. Consequently the idea foundered and, in any event, the successive Conservative government re-introduced the Duties in 1925.

The atmosphere between Cowley and Longbridge was further strained when the Wolseley company went into liquidation in 1926. Austin, who had strong commercial and emotional ties with the company, was keen to buy it, while Morris was also determined to add another marque to his growing empire (MG had emerged as a make in its own right in 1924) and to keep American interests at bay. Morris, Austin and their respective advisers attended what, in effect, was an auction for Wolseley, along with Julius Turner, General Motors' skilled negotiator. Morris finally purchased the company in the face of the expected opposition for £730,000 after informing Austin's Ernest Payton 'I am going just a bit further than you.'

Morris Minor

However, as his company grew, Morris, who became a baronet in 1929, began to lose his sense of direction in the late '20s. His troubles started with an unsuccessful bid for the small-car market in 1929. This challenge to the Austin Seven was the Morris Minor, which was powered by an 847cc overhead-camshaft Wolseley-derived engine. In retrospect, it can be seen to have been too sophisticated a power unit, both from a maintenance and manufacturing viewpoint. Therefore, a simpler side-valve version was introduced in 1931. In all, about 39,000 overhead camshaft Minors were built between 1929 and 1932, compared with 91,676 Austin Sevens produced over the same period.

This overhead camshaft Morris Minor saloon was available in 1930/31. But the model never sold in the numbers expected so a side valve engined version was hastily introduced in 1931

Ford of Britain

Then a major challenge to the small-car sector of the market came, surprisingly enough, from Ford. Hitherto, as we have seen, the British company had produced anglicized versions of the Model T. But their £120 Eight of 1932 with its stylish Detroit-designed bodywork and synchromesh on second and top gears was deliberately designed for the European market. By 1934, it had carved a massive 54 per cent of the 8HP range and it gave Ford a grip on the British market it has never relinquished.

The Ford Motor Company (England) Ltd., was founded in 1911 and the Model T was assembled at Trafford Park, Manchester, which was Britain's first trading estate, and conveniently bounded by the Manchester Ship and Bridgewater canals. The impact of the 'horsepower tax' has already been chronicled and in 1927 the Model T was replaced in Britain by the AF, a small-bored version of the American Model A, designed to circumvent the horsepower penalty. But in 1928 sales were a mere 6224 cars and commercials, a far cry from the heady days of 1919 when the Model T dominated British car sales.

The Manchester factory had been used previously for the production of tram bodies and was deemed an unsatisfactory manufacturing base by Detroit, so in 1923 Henry Ford's son Edsel and his brother-in-law, Ernest C. Kanzler, visited the marsh lands and rubbish tips of the Thames estuary at Dagenham, Essex. The choice of this site for the new Ford works was announced the following year and there were sceptics who believed that the boggy ground would be totally unsuitable for a massive manufacturing

plant. But the Thames-side position was a major plus giving quick and easy access to shipping from Europe and America. Ford had always insisted that his factories have deep water facilities on the lines of his mighty plant on the Rouge River, Detroit, which was the world's largest car manufacturing complex when it went into full production in the late '20s. He had refused to sanction a plant at Rotterdam, even after the foundations had been laid, because the docking facility was a kilometre away and Amsterdam was chosen instead. So Dagenham it was and in May 1929 Edsel Ford, accompanied by his young son Henry, cut the first sod at the site with a silver spade and the work began.

Above, in 1913 Ford was Europe's largest car maker with 6139 units and the following year a moving track assembly line was introduced at Trafford Park, another Ford first. Below, Percival Perry, who steered Ford's fortunes in Britain

By this time, Ford operations in Britain had been re-formed. The £7,000,000 Ford Motor Company was established in 1928 with Detroit controlling 60 per cent of the stock. The remainder was offered for sale on the British market, a tactic Ford usually adopted when creating offshoots abroad, so that the company would not be looked upon as wholly foreign. Ford also convinced Percival Perry, who had virtually created the British end of operations in pre-First World War days, and had left him in 1919, to return as chairman, and this he agreed to do.

Dagenham eventually went into production in October 1931 though it didn't become fully operational until the following year. Ford had intended it to be the 'Detroit of Europe' and the facilities were certainly impressive. It boasted its own power station, was the only car factory in Europe to produce its own steel and had a potential capacity of 200,000 vehicles a year.

Nice lines! Miss Spain 1932 with Ford's stylish Eight of the same year. Later a Spanish assembled version, the Forito, was produced from a Barcelona factory, an exercise cut short by the outbreak of the Spanish Civil War

However, the first Ford off the Dagenham production line was an AA truck, a reminder that this mighty works had very little else to build! The only way for the factory to pay its way was for Ford to produce a car specially tailored to the needs of the European market instead of relying on the large and thirsty American models, ill-suited to those depression years. This need had been pointed out back in 1923; in October 1931 Perry was in Detroit arguing this case and by February 1932 the European car was a reality! Codenamed Mercury, it had a simple 933cc side-valve engine with an 8HP RAC rating, compared with the AF's 14.9HP. The bodywork was designed by Eugene Gregorie, the company's chief stylist, and the car was displayed in February 1932, though by the time that the Eight went on sale in August he had made a number of changes to the style, which included increasing the slope on the radiator and giving the windscreen a fashionable 20-degree rake. Here was a car that Dagenham (and Ford's Cologne plant) could sell, as it was exactly geared to the times and the market. In fact Edsel Ford was so impressed by the appearance of the new Eight that the styling of the 1933 range of American Fords was *scaled up* from Gregorie's design. In Britain Ford had a best seller on his hands and in 1935, because of mounting opposition from Morris, the Eight's price was slashed to £100, the first time the market had seen a mass-produced saloon offered at this figure.

The Morris Eight, which appeared for the 1935 season, can be fairly described as a more refined version of the Ford product. Where Dagenham offered transverse leaf-springs, the Morris had half-elliptics and the Ford's mechanical brakes didn't really compare with the Lockheed hydraulics on the car from Cowley. The Morris was more expensive (£142 for the 4–door saloon with leather upholstery), but the public loved it and the Eight became the best-selling car of the decade with over 221,000 being produced. It revitalized Morris's flagging sales (in 1933 and 1934 Austin had produced more cars), but above all the Eight elevated Leonard Percy Lord's prestige within the British motor industry.

Top left, **Ford's Dagenham factory, photographed in October 1931. The quay and docking facilities are readily apparent and the plant's power station can be seen in the right foreground.** *Above*, **Len Lord with Lord Austin, pictured on a golf course in November 1938 after he had left Lord Nuffield (*Right*). They are shown with the happy owner of the 100,000th Morris Eight, built on June 30, 1936**

Leonard Lord

Len Lord (1896–1967), as he was always known, was a plain-speaking, single-minded individual and his maxim that 'if the door isn't open then you kick it open' probably says more than any of my words can. But, says his former colleague Miles Thomas, 'his apparent rudeness was a protective mechanism . . . like many men with a brusque unyielding exterior he was shy at heart'.* Lord was educated at Bablake School, Coventry, served an apprenticeship with Courtaulds and during the First World War worked at the Coventry Ordnance Works. He later joined the Daimler factory and in 1922 he moved to Hotchkiss's Coventry works, where, it will be recalled, engines for William Morris's fast-selling 'Bullnose' models were being produced. Morris bought the business in 1923 and Lord left his drawing office job, having been promoted to machine-tool engineer. At that time enormous strides were being made in engine production, master-minded by general manager Frank Woollard. From an output of 300 engines a week in 1923, production soared to 1200 by the end of 1924. By that date the factory boasted 786 machines laid out for flow-line production with an Archdale automatic transfer machine able to deliver a fully-machined and drilled Morris cylinder block every 4 minutes. Unfortunately it wasn't possible to

Out on a Wing, Michael Joseph, 1964

produce the rest of the 'Bullnose's' components at a similar rate, but it says much for Woollard's methods that they weren't adopted by the American industry until the Second World War and from then on by the rest of the motor manufacturing world. Lord was closely involved with these developments, both in the purchase and design of advanced machine tools. It was this experience gained at Morris Engines that gave him a commanding knowledge of the specialist business of producing and assembling large quantities of interchangeable components economically. He was soon to gain the reputation of being the country's finest production engineer.

In 1927 he was on the move again, this time to the Wolseley company, that had just been absorbed into Morris's growing empire. There, he was responsible for putting the overhead camshaft Morris Minor engine into production and, when that model didn't prove a success, he quickly created a side-valve replacement. Morris was clearly impressed with Lord's track record and, with a background of falling sales, he appointed him managing director of Morris Motors in 1933. Len Lord went to Cowley and certainly made things hum and as Miles Thomas was to recount: 'He walked roughly over the toes of anyone who got in his way.'* Lord was determined to re-equip Cowley and turn it into Europe's most efficient automobile plant capable of producing over 100,000 vehicles a year. A moving-track assembly line was laid down in 1934, something that most of the major European car producers had possessed since the early '20s. The first stage alone of this rejuvenation cost £300,000, but the results were an unqualified success and, with the appearance of the Morris Eight, Nuffield's star was again in the ascendancy. (Sir William became Lord Nuffield in 1934.) Lord also began the difficult task of bringing some sense of order to the rambling combine that consisted of a hotch potch of factories scattered around Oxford and the Midlands. The emphasis was on integration and the MG, Wolseley and Morris Commercial companies, hitherto privately owned by Nuffield, were, in 1935, drawn into the corporate fold. Unfortunately, Lord clashed with his employer over the prickly business of his share in the company's profits and slammed out of Morris Motors. Although Nuffield later asked him to administer a trust fund instituted to aid distressed areas, it could only be a matter of time before Lord again became involved in the motor industry. Soon after the break, Lord was sitting in Miles Thomas's office at Wolseley (where Thomas was general manager) and 'suddenly blowing the ash off the inevitable cigarette between his lips . . . said, "Tommy, I'm going to take that business at Cowley apart brick by bloody brick."'* Later events were to prove his prediction tragically accurate because in 1938 Lord joined Nuffield's arch rival at Longbridge. At 42 years of age he became Lord Austin's 'heir apparent'.

Len Lord made his mark on Cowley with a sweeping modernisation plan, begun in 1934, of which this moving track assembly line was a notable part. Compare the basic similarity of the layout with Ford's Trafford Park plant of 20 years previously on page 26. The Ten-Four model is seen under assembly

Vauxhall, by General Motors

Yet a further challenge for the buoyant motor markets of the '30s came from the other pillar of American industry in the shape of General Motors' ownership of Vauxhall. In 1925 the Corporation had set their sights on

**Ibid*

Austin and if Sir Herbert had had his way they would probably have succeeded, but he was outvoted by his fellow directors. Instead, the same year, they took over the Luton-based company which had produced a mere 1388 cars during the twelve-months. Later, in 1926, they also made an unsuccessful £11,000,000 bid for Morris Motors, while their attempts to take over Wolseley were also thwarted by the patriotic Morris.

During the '20s, General Motors in America, engaged in a furious sales battle with Henry Ford's Model T, was being welded into an efficient and aggressive conglomerate by Alfred Sloan Jnr. General Motors had been created by William Durant, a hard-selling carriage-maker from Flint, Michigan who purchased the ailing Buick company in 1904 and the Corporation was incorporated in 1908, by which time it embraced the Cadillac, Oldsmobile and Oakland marques, plus a host of other concerns, some of doubtful commercial worth. Durant soon found himself in financial difficulty and, in 1910, a bankers' syndicate took control, though he still remained a director. However, Durant later regained control and brought

A major milestone: Britain's first mass produced monocoque, the Vauxhall Ten-Four of 1938. This obviated the need for a separate chassis but sub frames were employed front and rear. In the post war years the manufacturers of pressed steel bodies played an increasingly crucial role in the car company's fortunes

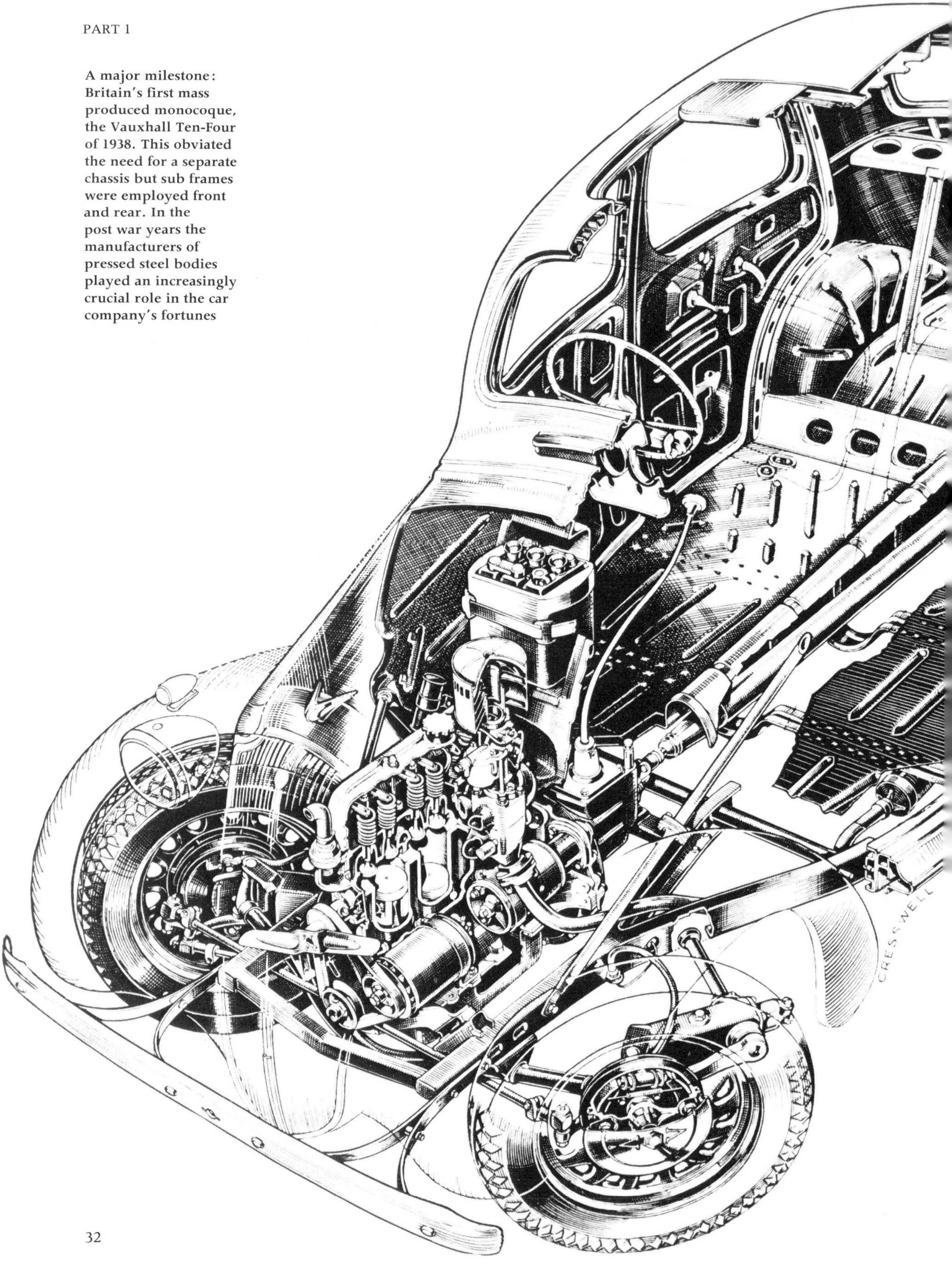

Chevrolet into the Corporation in 1916. It soon became apparent though, following another financial panic, that he was quite incapable of controlling such an unwieldy empire and, in 1923, Alfred Sloan (1875–1966) took over as President. Sloan, a graduate of the prestigious Massachusetts Institute of Technology, had run the Hyatt Bearing Company until it was absorbed by General Motors and he set about completely re-organizing the Corporation and decentralizing its activities. Each division was made a self-contained autonomous unit operating within the framework of the Corporation's policies. Sloan's management structure was founded on the belief that it was impossible for one man to run an organization the size of General Motors. This viewpoint was diametrically opposed to Henry Ford's dictatorial regime. But Sloan built up an able and efficient team and the attack on the Model T's supremacy was spearheaded by Chevrolet. In 1927 General Motors finally overhauled Ford in production terms and, although Henry managed to inch ahead in 1929–1930, the Corporation, with its handful of marques, having got in front, stayed there. Sloan's blueprint for a decentralized management structure has since been copied by large companies throughout the world even, after Henry's death, by Ford!

Alfred Sloan, President of General Motors 1923–1936, who finally stepped down as board chairman in 1956. Sloan's administrative genius turned GM into the world's largest and most profitable car company

This, then was the highly professional organization that took control of Vauxhall, although Sloan later admitted that he 'looked upon it as a kind of experiment in overseas manufacture'.* The company lost money for the first few years after the GM takeover, though an entry into the commercial vehicle market in 1931 looked more promising. Car sales, however, were still disappointing and in 1932 Sloan appointed a committee under Albert Bradley, his vice president of finance, who recommended that the existing range of cars be discontinued and that a smaller 6-cylinder be introduced and followed by a new four. The marque soon benefited from the vast resources of GM technology. Overhead-valve engines had appeared as a matter of course and the 1932 Vauxhall Cadet was equipped with a synchromesh gearbox, a 'first' for a British-built car but a Cadillac feature since 1928. Dubonnet independent front suspension arrived with the Light Six of 1935 and Vauxhall were the first manufacturer of the Big Six to adopt the feature.

A major milestone in production techniques came with the Vauxhall Ten-Four produced for the 1938 season. It boasted monocoque bodywork, the first mass-produced British car to be so constructed, the process having already been applied to the 1935 Opel Olympia from GM's other European satellite. Vauxhall established their own body plant at Luton known as the 'million pound shop' because the process was so expensive. The advantages were that the car's chassis frame could be dispensed with, thus saving weight, and a far more rigid body structure was achieved than had hitherto been possible. This latter feature was a major plus with the increasing popularity of independent front suspension. The disadvantage of the system though was that the manufacturer's tooling costs were increased and long production runs were vital in order to recoup expenditure. And for the car's owner, in the long term, the monocoque, with its many load-bearing pressings, was more susceptible to the rigours of corrosion.

By 1938, Vauxhall were fourth in manufacturing output amongst the Big Six and Bedford commercial vehicle sales were buoyant. Sloan's 'experiment' was bearing fruit.

The Rootes brothers, Billy (left) and Reginald pictured on board the *Queen Elizabeth* on return from America, winter 1946

Rootes Group

The remaining members of the Big Six were the Rootes Group and the Standard Motor Company. Both firms succeeded largely by building on already respected or established marque names, transforming and gearing their images to the populist demands of the 1930s. But, while Standard was run by the ruthless and quixotic John Black, the Rootes brothers laid down a filial management dynasty that continued until the Group's engulfment by Chrysler in the 1960s.

The Rootes story starts in the somewhat unlikely surroundings of Station Road, Hawkhurst, Kent, when, in 1898, William Rootes established a bicycle shop that soon embraced motor sales and repair. His eldest son, also William (1894–1964), after education at Cranbrook School, Kent, followed his father

**My Years with General Motors*, Sidgwick & Jackson, 1963.

into the motor trade, moving to Coventry to become an 'engineering pupil' as he later put it, with the Singer car company. After the outbreak of the First World War he joined the Royal Naval Volunteer Reserve as a lieutenant, but 2 years later, in 1917, he transferred to the Royal Naval Air Service and to the aero-engines that went with it. The same year he and his younger brother Reginald (1896–1979), established Rootes Ltd., a garage in Maidstone, Kent, the first link in a chain that grew so rapidly that, by 1926, the duo were running the largest car distribution business in the country.

Perhaps the reasons for this extraordinary growth lay within the complementary nature of the two brothers' personalities. For, while Billy, as he was always known, was a dynamic, thrusting salesman, Reginald, who had been a temporary civil servant at the Admiralty during the war, provided a perfect foil, offering restraint laced with financial acumen to direct this energy and imagination along worthwhile channels. If he was the engine of the Rootes combine, as Billy often remarked, then Reginald provided the steering and brakes.

This expansion was achieved by a deliberate policy of growth, coupled by financial integration and take-over of existing companies. In 1924 came an injection of Rootes' interests into George Heath's Birmingham garages and later involvements with Warwick Wright in London and Tom Garner of Manchester. In 1925 Rootes gained the controlling interest in coachbuilders Thrupp and Maberly of London, whose origins reached back to 1790. By this time Rootes Ltd. had established garages in Rochester, Kent, and London's Long Acre and in 1926 prestigious showrooms and headquarters were opened at Devonshire House, Piccadilly, opposite the Ritz Hotel.

But Billy Rootes's sights were clearly set on the manufacturing side of the business, with the products then being sold through their large chain of distributors. However, it was too late in the day to establish their own marque name, so the answer was to purchase an existing manufacturer. Initially they tried to take over the Wolverhampton-based Clyno company, whose products they handled, but when this failed in 1928 they took a substantial financial interest in the Humber Motor Company. One of Britain's older car manufacturers, Humber had been in the motor business since 1898. During the '20s they consolidated their reputation for producing good-quality, soundly-designed cars and, in 1926, they expanded into the commercial vehicle market by taking over the Luton-based Commer Cars. Having already taken an interest in Humber, the Rootes brothers looked around for another business to merge with it and, at the suggestion of A. C. Armstrong of *The Motor*, the Hillman Motor Company, who had been building cars since 1907, and were their next-door neighbours in Humber Road, Coventry, were taken over. (Armstrong was later presented with a handsome Asprey gold cigarette case by the brothers for his idea.)

Rootes' involvement in the Humber–Commer–Hillman combine grew to a complete take-over in 1932, the year that saw the creation of the Rootes Group. Rationalization was the order of the day and the 10 HP Hillman Minx, that went on sale in 1932 at £159, was well geared to the needs of the times. A further expansion of their commercial vehicle interests came in 1934 with the absorption of Karrier Motors of Huddersfield, production being subsequently transferred to Luton. The following year, the illustrious Sunbeam and Talbot companies were purchased, as a result of the collapse of the Anglo–French Sunbeam–Talbot–Darracq alliance. Sunbeam's Wolverhampton factory was sold off, though production continued throughout the '30s at Talbot's London premises. But individuality was banished and these finely-engineered cars became largely badge-engineered variations on the Humber–Hillman theme, the two makes being merged to create the Sunbeam–Talbot marque in 1938.

That distinctive look: the Standard de Luxe Flying Nine of 1938 that cost £162

Although the Rootes Group already employed the services of the Pressed Steel Company for many of their body panels, a desire for some autonomy in this field led to the purchase of British Light Steel Pressings of Acton in 1937. By the end of the '30s Billy and Reginald Rootes could look back on a decade of continual and uninterrupted growth. But, apart from expected competition, they also experienced a challenge from the products of the Standard Motor Company, which ironically arose from their involvement in the affairs of the Hillman company in 1928, where one John Paul Black was joint managing director.

Standard Motor Company

John Black (1895–1965) was born at Kingston-on-Thames, Surrey, and his legal ambitions drew him into the world of patent law. Inevitably this brought him into contact with many of the car manufacturers who were establishing businesses in those pre-First World War days. He was involved In the specifications of H. F. S. Morgan's ingenious 3–wheeler from Malvern and with the makers of the G.W.K. friction-drive car from Maidenhead. Black joined the Royal Naval Volunteer Reserve during the First World War and was later transferred to the Tank Corps in France. He was promoted to the rank of Captain, retaining the title, as was the fashion of the day, during the inter-war years. Black was demobilized in 1919, joined the Hillman company and by the late '20s he was running the concern, along with Spencer Wilks. Both had married Miss Hillmans and neither relished the Rootes' intrusion, so Wilks joined the ailing Rover company and 34-year-old Black accepted an invitation from Reginald Walter Maudslay, the Standard Motor Company's founder, to join them in 1929 as general manager.

Sir John Black, Standard's controversial managing director from 1934 until 1954, pictured at his office desk

Standard had been in the car business since 1903. The firm prospered until the mid-'20s, but no dividend was paid in 1927, and in 1929 losses had spiralled to around £120,000. Black soon re-invigorated the company, was appointed joint managing director in 1933, and took over fully the following year when Maudslay died at 64 years of age. Dividend payments began again in 1932 and reached an all time high in 1939. In 1930 Standard had built around 7000 cars and 9 years later the annual production figure stood at 50,700. At the end of the '20s the company had stood on the manufacturing sidelines, but by 1939 it was a well-established member of the Big Six.

This was achieved by a sweeping re-organization of the Canley factory and Black's determination to give the Standard products their own instantly recognizable indentity. Sales were already on the up and up when, in 1935, the Flying Standard range with its distinctive fast-back look appeared. Production soared and the company's best-selling year was 1939, a momentum unfortunately cut short by the outbreak of the Second World War. All this progress was achieved by Black's dictatorial and ruthless style of management, where fear replaced respect and the threat of dismissal was ever-present for those executives who dared to challenge this formidable personality.

It will therefore be seen that Standard, Rootes and Vauxhall made up the second division of the Big Six behind Morris, Austin and Ford. But, while Vauxhall remained financially secure beneath the mighty General Motors umbrella, Rootes and Standard, as comparatively late starters, were also potentially the most vulnerable members of the motor manufacturing league.

World trend setter: the *Traction Avant* Citroën of 1934. This was the first official photograph of the new Citroën front wheel drive range and shows a prototype 7CV. Small changes were subsequently made to the rear wings, roof line and bonnet handles. The bodywork is of monocoque construction, torsion bar suspension is employed and the engine is an overhead valve, wet liner unit. The *Traction Avant* remained in production until 1957 and many of its features were copied all over the world. In Britain, for instance, torsion bar suspension was widely adopted on many cars in the early postwar years. The model also fired Alec Issigonis's interest in front wheel drive

The state of the industry: 1939

As we have seen, government involvement in the affairs of the British motor industry had been mainly confined to the introduction of McKenna Duties in 1915 and the arrival of the 1921 'horsepower tax'. However, both parties were drawn closer together in 1936 when five of the car companies, Austin, Daimler, Rootes, Rover and Standard were asked to man, build and equip 'shadow' factories at public expense. The idea had been mooted in 1935 and was a recognition by the Air Staff that war with Germany was inevitable. The intention was to maximize production of the Bristol aero-engine and this was where the car manufacturers came in. As Sir Hugh Dowding, then head of the Supply and Research department of the Air Ministry, told Bristol's Roy Fedden in September 1935: 'We have called the technique "shadowing" because it aims to create a second image of the original.' In this instance the 'original' was the Filton works of the Bristol Aeroplane Company. Response from the industry was, in the first instance, lukewarm, but then Billy Rootes threw his weight behind the idea and this really set the ball rolling. So the first shadow factories were established, initially geared to Bristol Mercury and Pegasus aero-engine production, and were mostly built very close to the parent car company's own works.

These vast new factories and the mechanism of their creation is important to our story. It meant that, after the war, they could be taken over by the industry for vehicle production in the car-hungry '40s and '50s. But on a broader basis they forged a link between the government and the motor industry which was consolidated by war. The one would never again be independent of the other.

At the outbreak of the Second World War in 1939 the British motor industry was, apparently, in a healthy and ebullient state. It was one of the growth centres of the pre-war years, along with the chemical and electrical industries, and was Europe's largest in terms of output.

But the American challenge, perhaps exemplified by the Ford Eight of 1932 from that highly integrated Dagenham plant, was re-awakened in the '30s. It increased steadily over the decades, led to Austin and Morris joining forces in 1952 and eventually to the creation of the British Leyland Motor Corporation in 1968.

Many of the weaknesses and failures of the postwar years within the British-owned sector are clearly displayed in the '30s: the products themselves, the markets they served and in the backgrounds and personalities of those who shaped the industry, William Morris and Herbert Austin.

Let us start by considering the products. Just how did they compare in design with those of our continental neighbours? The German industry was next in size to the British one, though the rigours of depression resulted in a smaller number of firms, and anyway such amalgamations were established German practice. In 1937 the top-selling marques were Opel, DKW, Adler and Mercedes-Benz in that order. In every instance, from the cheapest to the most expensive, independent front suspension was standardized while 2 of the firms (DKW and Adler) offered front-wheel drive models. Bodywork showed, in some instances, the results of wind tunnel research, which was a fuel-saving plus on the new *Die Reichsautobahnen*, the first of these motorways being opened in 1936.

In France in the '30s Citroën was the best-selling marque, narrowly in front of their rivals, Renault. Design was less advanced than that of German, but Citroën's *Traction Avant* model was a glorious exception. This low-built monocoque front-wheel-drive car of 1934 was powered by a 4–cylinder overhead-valve wet-liner engine, employed torsion-bar suspension which was independent at the front and (from 1936) was fitted with rack and pinion steering. Unfortunately Andre Citroën overreached himself in the production of this world trend-setter and lost control of his company to Michelin, his largest creditor. With such advanced thinking in both France and Germany it is not surprising that, by 1939, one in four continental cars had front-wheel drive.

Car production in Italy virtually meant Fiat, and their 500 of 1936 again represented a progressive point of view with its tiny forward-mounted engine and aerodynamic bodywork. It was obviously in the spirit of the Austin Seven and one distinguished member of the Institution of Automobile Engineers went so far as to point out that this was the sort of car that Longbridge should have been building rather than relying on the faithful Seven, which had been designed in 1922.

In Britain front-wheel drive had been attempted by only 2 firms, Alvis and BSA, but neither efforts were commercial successes and sales were small. Independent front suspension was employed by only half the members of the Big Six, the American-owned Vauxhall company, Rootes and Standard, while Vauxhall also initiated the production of monocoque bodywork to the British motor industry.

Of course, a strong body of opinion would rightly argue that innovation should be a gradual process and too revolutionary a car could bankrupt a company. Citroën's collapse in 1935 no doubt caused much head shaking at Longbridge and Cowley, but the point I am making is that British mass-produced cars were designed as they were because things had *always been done that way*; imitation rather than innovation was the order of the day. Things were little better in the small-production British sports car market and one looks in vain for a machine of the stature and sophistication of, say, the German BMW 328.

Laurence Pomeroy, famed technical editor of *The Motor*, in summing up the Austin and Morris products in the postwar years, put it this way: 'They were designed by successive generations of engineers, who must have had a common sympathy with the sentiments expressed by the Duke of Cambridge, who said "all change at any time is to be utterly deprecated".'

The reason for this state of affairs lies in the fact that both companies' designers mirrored the attitudes of their proprietors. Morris and Austin were essentially self-taught men, cast in the Victorian mould, who triumphed on a basis of drive, intuition and sheer hard work, but were possessed of little or no technical knowledge. It should be remembered that in Britain, elementary education only became compulsory in 1880 while secondary teaching did not become a national responsibility until 1902. What little technical education there was came too late and was continually starved of finance. Therefore, British nineteenth-century engineering traditions were based on a strong practical footing, with a high regard for the individual product.

By the end of the nineteenth century France and Germany had benefited by almost 100 years of compulsory schooling, the latter country possessing the finest system of technical education in the world, which in turn gave birth to the motor car in 1885. Therefore, with a few exceptions, of which the Lanchester was the most notable, British cars were essentially copies of continental and, later, American designs, a state of affairs that continued for the following 60 or so years.

But if good scholastic opportunities were lacking in Britain, there was unfortunately little incentive for those who had benefited from a good education to enter the engineering profession, because such a pursuit did not accord with the social attitudes of the day. The government and ruling classes, who set the social pace, were almost exclusively products of the public schools and old universities, where the teaching and environment was steeped in the Classics, which had little connection with the technological needs of the day. This doctrine, with a few exceptions, was anti-scientific and technical, such subjects not being within the orbit of a Christian gentleman having the taint of trade and all that that entailed. As Sir Alexander Fleck reminds us: 'A bias remained against trade and industry, a bias that derived from the Platonic tradition which regarded money and mechanical labour as vulgar and degrading.'*

I can illustrate this attitude in no better way than by drawing on the formative years of 2 outstanding figures of the British industrial scene: Sir Miles Thomas (later Lord Thomas of Remenham), who became vice chairman of the Nuffield Organization, and was later a highly successful chairman of BOAC; the other is Sir Roy Fedden, the driving force of Bristol aero-engine design between the wars. Both were products of a public school education; in Thomas's case Bromsgrove and in Fedden's, Clifton College. Thomas, whose father died the year after he was born, had shown a wish to enter the engineering industry, but his mother 'wanted me to go in for something which she called more "respectable". . . . She was thinking in terms of the Church, Law, or, particularly, of medicine. I was thinking in terms of something more dynamic, of movement and of metal'.† Fortunately, Thomas got his way, aided by his headmaster, R. G. Routh, who supported young Miles in his objective and he entered the Birmingham engineering firm of Bellis and Morcom as a premium pupil.

Similarly, Roy Fedden was determined to enter the motor trade, though his headmaster, Canon Glazebrook, favoured Sandhurst and a military career. Nevertheless, Fedden was resolved to pursue his ambition, though his father was profoundly shocked, but he later told young Roy: 'If you are so keen on engineering then that is what you must do.' Perhaps even more significant though, was the reaction of the mother of a young girl friend who could hardly believe her ears. As Bill Gunston recounts in Fedden's biography: 'In Roy's own words, pronounced over seventy years later, "she was convinced that I was going to be some kind of plumber, dirty and brutish. She no longer wanted me in her house and suggested that, if I did come, presumably I would go round to the *back door*".‡

These experiences do suggest that those products of the public schools

**A History of Technology, Volume V*, OUP, 1979
†*Out on a Wing*
‡*By Jupiter*, Royal Aeronautical Society, 1978

who did enter the engineering and motor trades were swimming against the social current and one can only speculate on the number of homes where hopes of an engineering career were dashed and a more socially acceptable pursuit taken up. It is a viewpoint that persists *to this very day*, and consequently the British engineer has never enjoyed the same social status as his continental or American counterpart. A survey for the 1980 Finniston Report on the state of the engineering professions and manufacturing industries showed that 68 per cent of the British public thought that an engineer 'was someone doing manual work', probably connected with machinery, while only 13 per cent associated the title with design or research at a professional level.

With this background of a late educational start and strong social pressures against engineering it is difficult to argue with Sir Harry Ricardo's opinion that, before the First World War, with a few exceptions, 'the design of road vehicles had been in the hands of cycle makers, superb mechanics well versed in the arts of light mechanical design but, for the most part, abysmally ignorant of thermodynamics, or of the many other factors upon which the performances of their engines depended'.*

No major figure of the pre-war motor industry more personifies this essentially empirical approach to design than Herbert Austin. While he possessed a natural aptitude for mechanics his cautious approach to car design, largely through a lack of knowledge, reveals itself again and again by his leaning heavily on other engineers' creations. His first design for Wolseley was largely a copy, though it must be admitted an improvement on the contemporary Léon Bollée. But when he went into business on his own account, as that shrewd observer of motoring history, the late Anthony Bird, has pointed out: 'Austin models from 1906 onwards therefore were unashamedly copied from other sources . . .' and with one exception 'clung to the practice of using separate cylinders long after most of his contemporaries had abandoned the idea'.†

Stanley Edge, who created the diminutive 4-cylinder engine that made the Austin Seven such an outstanding success, had an opportunity for studying Austin's approach to design at first hand. And again the intention was to copy, in this instance the Rover Eight, an existing model. Fortunately, as recounted, Edge got his way but 'I never saw him make a design calculation nor did he give me the results of calculations. I knew he was keen on estimate weights and working out costs, whether correctly or not I do not know.' And again: 'When I said the bore and stroke [of the engine] will give 696cc to Austin he simply said "will it?". Things like inertia of the reciprocating parts, the same for the valves to settle, spring rates or inlet and exhaust valve speeds he just wasn't bothered with.'

This innate conservatism was reflected by the backward design of the bodywork and mechanics of the Longbridge products of the '30s, which changed little in layout to the models of the previous decade. While some of his competitors offered overhead valves, independent front suspension and hydraulic brakes, Austin still clung to the side-valve engine, cart springing

**Memories and Machines*, Constable, 1968
†*The Veteran Motor Car Pocketbook*, Batsford, 1963

and mechanical braking. Not surprisingly, his market share dropped from just over 37 per cent in 1929 to around 24 per cent 10 years later.

Morris, who was 11 years younger than Austin and less steeped in practical engineering traditions, was more of an innovator than his Longbridge counterpart. He, of course, achieved enormous success in the '20s, but towards the end of the decade he began to absent himself from the company's executive meetings by taking long sea trips to Australia. Although he maintained that he left the running of Morris Motors to others, if a vital decision was taken in his absence it infuriated him and this dictatorial approach to running his empire meant that he argued and then fell out with practically all his chief executives. His production genius, Frank Woollard, departed in 1932, Len Lord went to Austin and finally Sir Miles Thomas left after an argument in 1947.

Perhaps, at heart, Morris was really a small businessman. Graham Robson has described him as 'the sort of man who worried about lights being left burning in empty offices';* his dislike for large factories was obvious to his colleagues and he seemed far more at home in small manufacturing units. His business grew and then expanded further by acquisition of an assortment of factories. Perhaps he preferred his scattered empire the way it was, but the haphazard nature of its growth has presented operating problems to this very day. A greater contrast to the highly integrated purpose-built Ford Motor Company's Dagenham plant is difficult to imagine.

Although Austin and Morris's relations had been impaired by the Wolseley take-over battle of 1927 both men were in accord in their attitude to university graduates as is sometimes the way with self-made men. Austin went so far as to dogmatize 'the University mind is a hindrance rather than a help' and Morris went even further. As one of his biographers, William Jackson, points out, Morris even sought out some of his university-educated employees and sacked them. Len Lord supported this general viewpoint and it is not, therefore, surprising to find that BMC did not begin a policy of graduate recruitment until 1963 after Lord's retirement. The Ford approach is again strikingly different. Although in pre-war days, their chief executive, Percival Perry, shared similar sentiments to Austin and Morris, the postwar attitude was very different. As Graham Turner points out in *The Car Makers*, published in 1963: 'Ford hires hundreds of university graduates — classicists, linguists, historians — to do the job, to analyse and compute costs, and hence profit'.

Yet another weakness of both the Austin and Morris companies was their enormous model diversity. In 1919 each firm started off with one basic design. Yet, in 1934 Austin was offering a choice of 44 separate models based on nine alternative chassis and, if varying colour schemes are taken into account, this amounts to 333 different lines. Similarly, in 1933, Morris was offering, Eight, Ten, Twelve, Fourteen, Sixteen, Eighteen and Twenty HP cars. Although both companies did cut down on this plethora of models and there was undoubtedly some interchangeability between ranges, by 1938 as *The Economist* was pointing out that the industry was producing 'no less

**Motoring in the 30s*, Patrick Stephens, 1979

than 40 different types of engine and even larger numbers of chassis and body types'. Across the Atlantic, in America, the Big Three (General Motors, Ford and Chrysler), which produced close on 4,000,000 cars in 1937, actually had *fewer* models in production than our Big Six manufacturers with around a tenth of the output. We therefore find this American thinking reflected in the products of Dagenham and Luton with far fewer models and a greater interchangeability of parts.

There is a final point to be made about the British pre-war car market. In 1938, for example, about 341,000 cars were produced, and of these 85 per cent cent were sold on the home market, which was protected by McKenna Duties. Of the remaining 44,000 exported, 75 per cent were sold in Imperial Preference markets (Australia, New Zealand etc.). Therefore, 97 per cent of cars manufactured in Britain in 1938 were sold in *protected* markets where the vehicles enjoyed a significant price advantage over their competitors. Britain was not to feel the real effects of foreign competition until the 1950s.

During the Second World War the motor industry responded magnificently to the demands of the fighting services. The original shadow factories, already in production, were soon followed up by others. The industry's output varied from aero-engines, bomber and fighter aircraft to motor and armoured vehicles and tanks.

Austin, the Rootes Group and Standard produced aeroplanes, Ford tractors, trucks and lorries as well as Rolls–Royce Merlin engines from a specially-equipped Manchester factory. Morris repaired aircraft, produced mines, tanks and torpedoes, while Vauxhall designed and developed the Churchill tank in record time. And almost everyone, from Austin and Morris to Vauxhall seem to have produced jerrycans by the million!

These efforts did not pass unrewarded. Ford's Patrick Hennessy, seconded to the Ministry of Aircraft Production, was knighted in 1941, Billy Rootes became Sir William the following year and John Black and Miles Thomas received similar accolades in 1943.

But the demands and potential markets of the postwar years were not ignored. Fire-watching provided an ideal opportunity for engineers to get together to discuss future designs and two of the most significant, the Jaguar XK engine and the Morris Minor emerged from such sessions. Fortunately, the Society of Motor Manufacturers and Traders persuaded the government to allow firms to push ahead with prototype work for new designs in 1943 and by the following year no less than 56 companies were reported to have such projects under way.

The Healey was one of the new makes to emerge in the early postwar years. Pictured at the Warwick factory collecting his Healey Silverstone in 1950 is Italian driver Aldo Bassi (second left), the Healey agent for Brescia. At the wheel is Count Maggi with Donald Healey in the passenger seat. Behind them, left to right, are Donald's son Geoffrey, Renzo Castegneto and A. C. 'Sammy' Sampietro (who played a formative role in the Healey project). Castegneto and Count Maggi were organisers of the Mille Miglia

Postwar problems

At the close of the Second World War in 1945, a trickle of cars began leaving Britain's factories, a mere 17,000 or so vehicles being produced. The pace quickened the following year when around 220,000 cars were made, which was around the 1933 level. Output continued to rise but the real indication that things were getting back to normal came in 1948 when the first Motor

Show for 10 years was held. No less than half a million people thronged to London's Earls Court that October, which was almost double the 1938 attendance, to see 33 British makes displayed, although of these the Big Six accounted for 13 marque names. Unfortunately there was little chance of buying a new car as that year 69 per cent of production was directed by the government to the export market. Nonetheless, Morris and Austin were vying for superiority as top car producer and visitors to the show were reminded that a Britisher, John Cobb, was the fastest man on earth as his sleek Railton Mobil Special was given a special stand on the ground floor of the exhibition hall.

The industry had remained largely intact during the decade since the previous show and, even if such names as BSA and British Salmson hadn't survived, then their absences were compensated for by three new makes: Allard, who had been building specials since 1934, the finely engineered Bristol and the chunky sporting Healey from Warwick. Changes were more apparent in the Carriage Work section. In 1938 26 coachbuilders had exhibited and 10 years later the figure had dropped to 18. For the first time Briggs Motor Bodies and Fisher and Ludlow displayed their pressed steel

MHP 102

Two makes destined not to survive the 'fifties. *Left*, Lanchester production ceased in 1956. A 1953 Leda, an export version of the 2-litre Fourteen is shown. *Below*, a 1953 Lea-Francis 14, the company's last year of manufacture. The marque was briefly revived in 1960 and has reappeared in 1980

***Overleaf*, high hopes — a Bristol 400 saloon at Filton in 1948 against a backcloth provided by the Bristol Brabazon. The aeroplane was completed the following year but scrapped in 1953 after completing 400 flying hours**

bodies, a reminder that quantity production was to be the keynote of the postwar years.

Things were very different at the 1967 show. The number of marque names had dropped to 30 and of these the Big Five, as they had become, accounted for 14. Ford had a winner on their hands with the Cortina saloon which was that year's top-selling model.

Austin and Morris had combined in 1952 to form the British Motor Corporation and in 1966 it took Jaguar under its wing (that company having absorbed Daimler in 1960) to form British Motor (Holdings). Standard–Triumph and Rover, and the latter had taken over the Alvis company in 1965, were now the property of the thrusting Leyland Motors while the ailing Rootes Group was being progressively absorbed by the American Chrysler Corporation.

A host of smaller companies had ceased production during those 19 years. Jowett's brave attempt with their advanced Javelin floundered in 1953 and Lea-Francis, with a generation of financial ups and downs behind them, faded from the scene in 1954 and was briefly revived in 1960. Singer succumbed to a Rootes take-over in 1955 while Lanchester was a casualty of the fading fortunes of the BSA combine in 1956. Small producers like Allard and Frazer Nash stopped building cars in 1960, the same year the last Armstrong Siddeley left Coventry. And Alvis built their final car just a few weeks before the 1967 show opened.

Gone were such illustrious coachbuilding names as Freestone and Webb, Gurney Nutting, Hooper, Thrupp and Maberly and Tickford. Vanden Plas had been elevated to a marque name by BMC, though Mulliner — safe under Rolls–Royce's ownership — Abbott and BSA Carbodies survived. However, the appearances of Bond, Fairthorpe, Lotus, Reliant and TVR were reminders that glass-fibre and resin had replaced aluminium and ash as the materials of the small-quantity car builder. But of these only Reliant, TVR and, above all, Lotus were destined for survival.

British production had roughly doubled between 1952 and 1958, a year which saw over 1,000,000 cars leave our factories for the first time. Even though the family saloon and new sports-car markets were expanding, in 1956 we ceded our second place in the world manufacturing stakes to West Germany, and by 1967 had dropped to fourth position behind the USA, West Germany and France. Imports for 1967 stood at around 93,000. How different from those days in 1946 when just 63 cars entered our country from abroad!

The situation that faced the British motor industry at the end of the Second World War was in complete contrast with the boom conditions that greeted the cessation of the first great conflict in 1918. Such had been the financial demands of war that, by the beginning of 1941, practically all the country's economic reserves had been exhausted. A buoyant balance of payments position pumping valuable foreign currency into the exchequer was therefore an economic cornerstone of Clement Attlee's postwar Labour government. Exports were the order of the day and consequently the motor

S.G.B

industry suffered from controls that it had never previously experienced in peace time. War-time constraints over steel production were perpetuated, with firms allotted tonnage commensurate with their export performance.

As we have seen, before the war the industry had been almost exclusively geared to the needs of the home market. Therefore, after 40 years of serving the British motorist, manufacturers had to look upon the world as their market place and exports boomed, though at the expense of home demand. With practically no competition from Germany, France or Italy, who were re-building their economies, British cars experienced little competition in foreign markets in the late '40s and early '50s. Such was the export achievement that, in 1950, Britain overhauled the USA as the world's largest exporter of motor vehicles.

In those early years, the markets were mainly to Australia and New Zealand and, to a lesser extent, America; exports to that continent didn't peak until the late '50s. MG and Jaguar sports-car sales to the USA, though initially small, grew rapidly, perhaps re-awakening a dormant market for open 2–seaters. In 1951 Jaguar recorded an all-time record for a British car company when it exported no less than 96 per cent of its total production. Therefore, the industry didn't really turn its attention to the home market until 1955. For instance, the 1937 level of new-car registrations in Britain was not approached until 1953 or exceeded until the following year. Not surprisingly, pre-war cars were commonplace in the booming '50s, and the home market's needs were not fully satisfied until the early years of the following decade.

Government controls of the industry didn't stop with the export drive. Purchase tax was introduced on cars in 1940 at the rate of 33.3 per cent but, on a more positive front, the 'horsepower tax' was finally abolished. Although the annual Road Fund Licence had been paid on the basis of £1 per RAC horsepower since 1921 it was reduced to 15 shillings in 1934, but increased again to 25 shillings in 1940. As will be remembered, it constrained manufacturers to produce narrow-bored long-stroke engines and during the war the Society of Motor Manufacturers and Traders, with this limitation in mind, suggested to the government that the annual tax be calculated on the basis of an engine's cubic capacity, instead of the RAC rating, and this system was duly implemented in January 1947 graduated in steps of £1 per 100cc. However, that spring the National Advisory Council for the Motor Manufacturing Industries reported their preference for a flat rate of tax as it was felt that the graduated system would be a disincentive to car makers to produce the larger vehicles required by the export market. As ever-increasing overseas trade was an integral part of the government's economic policy this recommendation was adopted and a new flat rate of £10 per car was introduced at the beginning of 1948, though this only applied initially to cars registered after January 1947. However, to offset the loss of revenue from large-capacity cars the rate of purchase tax for vehicles costing over £1000 was doubled to 66.6 per cent. Later, in 1951, all cars had to pay the new rate, a state of affairs that continued until 1955.

Yet further state pressure was applied to the car makers to limit their output to a single model, which was certainly at odds with pre-war fashion. Standard, for one, took up the Ministry of Supply's pleadings with their solitary Vanguard model of 1948 but, at the other extreme, by 1949 Austin was producing the most comprehensive range of cars in Europe! There were similar governmental demands for standardization of components. In the electrical field, for instance, just before the war there were 48 different sorts of dynamo, 32 starters, 12 coils, 18 batteries, 98 windscreen wipers and a staggering 133 different types of headlights available. Positive steps were soon under way to remedy this untidy state of affairs.

Austin and Morris merge

On the manufacturing front, Morris and Austin, the two market leaders, seemed destined for future competition. Longbridge, under Len Lord's aggressive direction (he had become chairman and managing director in 1945) was quicker off the mark than the Nuffield Organization and was being lavishly re-equipped for the years of peace. Meanwhile, at Dagenham, Sir Patrick Hennessy, who had worked for the Ford company since 1920, gradually assumed overall control, becoming managing director in 1948 and chairman in 1956, and skilfully guided the company's fortunes until he retired in 1968. Ford had re-started production in 1945 with their cheap, primitive, but tough pre-war range of cars characterized by 6-volt electrics, vacuum windscreen wipers that ceased to operate going up hill and transverse leaf-suspension that dated back in concept to the Model T of 1908! In 1946 Ford's market share stood at 14.4 per cent, but they grew faster than any of their contemporaries in those early postwar years and by 1951 their penetration had climbed to 18.1 per cent. While Vauxhall's growth hadn't been anything like as spectacular, the two American companies did present a serious threat to the British-owned sector. But there the problem was not so much with products but personalities.

In the face of this transatlantic challenge a merger between Austin and Morris seemed logical and it should be remembered that the discussions between the two companies in 1924 were held to combine against the might of the American motor industry. Although the problem then had been one of imports the threat in the late '40s was basically the same, though this time the 'enemy was within the gates'. And while Austin were on the up and up, 70-year-old Lord Nuffield was again endeavouring to rationalize his untidy empire. There was a spate of management sackings at the end of 1947, and Sir Miles Thomas, vice-chairman of the Nuffield Organization, and its most able executive resigned. It had been Thomas who had encouraged Alec Issigonis to pursue his war-time Mosquito car project that was to emerge as the beloved Morris Minor, but because of Nuffield's procrastination, the car's public appearance was delayed until 1948. 'Good god,' he told Issigonis, 'it looks like a poached egg — we can't make that.'

The barrier to reconciliation went back, of course, to Lord's row with

The export drive resulted in British cars reaching some unlikely locations. Here members of a Beaver indian tribe in Canada make the acquaintance of a Morris Minor

Nuffield in 1936 when he slammed out of Cowley muttering vengeance and then joined Austin, the tycoon's arch business rival. Although some mild co-operation between the two companies occurred in 1948 the stalemate wasn't broken until October 10th 1950, which was Nuffield's 73rd birthday. On that day Len Lord phoned his friend Carl Kingerlee, who was Nuffield's secretary, and asked him to pass on his best wishes to the peevish peer. Kingerlee responded that Nuffield was only in the next room, so why should Lord not speak to him directly? Initially Nuffield refused, but at Kingerlee's urging he then agreed to do so. The ice between the two men was cracked, if not thawed, and Lord soon followed up with a visit to Nuffield's house and then to Cowley. Both parties agreed on the need for amalgamation and the proposition was put to the respective boards of directors. However, Nuffield experienced strong hostility from Reginald Hanks, who had taken over from Miles Thomas, as he thought the proposed merger was biased in Austin's favour. Nuffield was swayed by the argument and telephoned Lord to tell him that the deal was off. Lord then informed Kingerlee that Austin had decided to go ahead with their new small car, the A30, which was designed to compete with the Morris Minor, entailing a duplication of resources that perhaps encapsulated the need for a merger. Communications subsided for about a year and then Lord made yet another overture. On this occasion he proved successful and Nuffield agreed to the proposition, regardless of what his board of directors might think.

The outcome of the decision was announced at a hastily-assembled press conference at Longbridge in November 1951. As a result of the merger the new company formed the largest manufacturing combine outside the USA. The British Motor Corporation formally came into existence in February of the following year, with Lord, the architect of the union, at its head. Thus the Austin, MG, Morris, Riley and Wolseley marques were combined, their market share in 1952 approaching 40 per cent, just about double that of Ford.

Small beginnings, the Morris Mosquito (a name suggested by Miles Thomas) pictured at Cowley in 1944. It later developed into the Morris Minor after Alec Issigonis had a prototype sawn in half and widened by four inches!

The Corporation's headquarters was based at Longbridge and from then on Nuffield took a back seat, becoming BMC's President, a purely honorary position, until his death in 1963. Unfortunately Lord's acerbic approach was such, that many of the Morris personnel felt humiliated and usurped by the Longbridge orientated hierarchy. In many respects the union just didn't gel, with each company retaining its individual loyalties, dealers and accounting systems. Joe Edwards, who became BMC's managing director in 1966, told Graham Turner for his crisp and sagacious *The Leyland Papers** that he considered that Lord's personality had set the merger back a decade.

On a more positive front a 3-year rationalization plan was immediately implemented. At the time of the merger the various companies produced a jumble of side- and overhead-valve and overhead-camshaft engines. These were soon reduced to three basic types. The two smaller units were designated A and B types and were existing Austin designs, while the larger and subsequent 6-cylinder C type was a Morris project. A similar rationalization approach was adopted for gearboxes and rear axles.

Another important innovation lay in the area of body design. Longbridge,

*Eyre and Spottiswoode, 1971

unfortunately, had little tradition for good styling. The models of the '30s were largely minor variations of the boxy theme of the previous decade while the immediate postwar cars were mostly scaled down versions of American shapes. Lord was certainly aware of these deficiencies and, prior to the creation of BMC, had asked the London office of industrial designer Raymond Loewy for a project that emerged as a good-looking two-door sports coupé. Loewy had already worked wonders for Studebaker in America, but the designs were never taken up. Morris fared a little better than the Longbridge products in appearance, but again transatlantic themes were all the rage. With the creation of the new corporation, body styles had to be rationalized, but there were, obviously, still products from both sides that had to be utilized and it wasn't until 1958 that the first truly corporate car emerged: the Austin A40 saloon.

In 1956, Lord, along with the Italian car makers, the American Nash Corporation and the French Peugeot company, retained the services of stylist Pinin Farina. This Italian styling house produced in the A40 a crisp, pleasing design which pioneered the 'two box' rather than the traditional 'three box' look, with which we are so familiar today by its apparent absence of a boot. In fact *Design* magazine went so far as to say that 'it is certainly the first BMC product which has undeniable "style"'. This was soon followed up by a rather less inspiring but larger four-door saloon in 1959 and, unlike the smaller car, was designed to carry the Austin, MG, Morris, Riley and Wolseley name plates. Badge engineering had arrived in earnest.

Ancient and modern; Ford products in the postwar years. *Right*, eight horse power meets one horse power, a 1946 Anglia with friend. *Below*, a switch from verticals to horizontals. Ford's first monocoques were the Consul and Zephyr models of 1951. This Consul employed a $1\frac{1}{2}$–litre four cylinder overhead valve engine and cost £531

Ford makes progress

Meanwhile Ford had not been idle. In 1950 Dagenham announced their new Consul and Zephyr range, the two cars being designed to employ the maximum number of common components. New 4– and 6–cylinder engines were employed, their larger bores and short strokes reflecting a new approach to design unencumbered by considerations of the old RAC horsepower rating. MacPherson strut independent front suspension was introduced to Britain on these models, while the monocoque bodywork was up to date though, not surprisingly, rather American in appearance. A similar theme was applied to the smaller 100E model of 1953, though in this instance a side-valve engine was retained. The company was building on its reputation of offering reliable, straightforward no-nonsense products at, above all, the right price. However, the archaic pre-war designed Popular continued in production up until 1959 and was the cheapest car on the British market.

In 1953 Ford took over Briggs Motor Bodies, the country's largest mass-producer of car bodies, who had supplied them since the beginning of their Dagenham operations in 1931. Sir Patrick Hennessy later spoke of the purchase as one of the most important events in Ford's postwar history as it gave them autonomy in the body-building field. The reverberations were felt throughout the entire British motor industry.

FMC 7

At Ford's request the Detroit-based Briggs Manufacturing Company had established its British subsidiary alongside the new Dagenham plant but to avoid total dependence on the industrial giant, Briggs had deliberately sought other contracts in the pre-war years, supplying bodies to such firms as Daimler, Riley and Standard. After the war, their largest customer, after Ford, was the Jowett company, with Briggs producing ready trimmed and painted shells for the new, and in many respects, outstanding Javelin model, from a Doncaster plant established during the war. Unfortunately, Jowett, a consistently undercapitalized concern, ran into deep financial problems in 1952 when Javelin sales slumped, probably because of the model's growing reputation for unreliability, and they asked Briggs to halt production. In the meantime, Hennessy was convinced that Chrysler, who was Briggs' largest American customer, would buy the parent company following the founder's death — and later events proved him right — 'and then control our body supply'. Therefore, between 1951 and 1953 he made numerous attempts to buy Briggs' British subsidiary and finally clinched the deal over the telephone to America early in 1953. But it also spelt the end of Jowett as a car manufacturer because they were unable to supply Briggs with convincing evidence of their financial resources. The ripples of the Ford take-over soon reached BMC and within 6 months the new corporation had followed Dagenham's example and snapped up Fisher and Ludlow of Birmingham, causing great concern to Standard, who relied on that company for much of their body supply.

American influence. Sir John Black with a 1947 Standard Vanguard, with styling inspired by a Plymouth saloon and destined for world markets. These Welsh hills proved an invaluable testing route

Standard–Triumph and Massey–Harris–Ferguson

Standard's controversial managing director, Sir John Black, had taken over the moribund Triumph company in 1945 with the intention of giving William Lyons's stylish Jaguar sports cars a run for their money. The same year he met up with Irish millionaire Harry Ferguson and obtained a highly lucrative licence to manufacture the Ferguson tractor in Britain. Black was a firm believer in a one-model car policy and, in 1948, all his other models were dropped and replaced by the new Vanguard saloon, its engine also being used in the Ferguson tractor. The single-model approach lasted for 5 years and 1953 saw the Vanguard joined by a new small car, the slightly austere Standard Eight. Also the Triumph marque really got into its stride with the appearance, in 1952, of what was to become the TR2, the first of a line of well-regarded and competitive sports cars.

Sir John's unpredictable, and almost schizophrenic, personality resulted in a boardroom revolution in 1954, his assistant, Alick Dick, taking over as managing director. Standard was destined to become the smallest member of the Big Five by the end of the decade and Dick believed that, if the company was to survive, then a merger with another manufacturer was essential. There were on/off talks with Rover with names like Allied or United Motors being mooted, but they eventually came to nothing and conversations with the Rootes Group suffered a similar fate. Standard's other problem was that

BMC, as a major competitor, now owned one of their principal sources of bodyshells, so Dick set out to ensure the company's independence in this crucial area. In 1954 he obtained an agreement with Mulliners of Birmingham, who already produced some of Standard's bodywork, for exclusive rights to their entire output, once they had cleared existing contracts to other customers. This rather upset Alvis, Aston Martin and Daimler, but the deal went through and 4 years later Standard purchased Mulliners Ltd. By this time the new Triumph Herald project was well under way and in 1959, the year of its announcement, Standard pulled out of the profitable Ferguson tractor enterprise. The money that the company received from Massey–Harris–Ferguson was used to buy another body-pressing concern, Hall Engineering of Liverpool, and also to build a new car assembly plant at nearby Speke. As the decade neared its end, Dick seemed to be inching his way forward to achieve at least one of his objectives.

Others in the Big Five

The remaining members of the Big Five, Rootes and Vauxhall largely retained their respective market shares, built up in the '30s, though neither could match Ford's postwar growth. The remainder of the market was made up by

DGW 29

Left, Like many manufacturers, Vauxhall offered their pre-war models in the postwar years. The Twelve de Luxe saloon was available until 1948

Up market Rootes. The Humber Hawk Mark VI saloon with a new 70bhp 2.3–litre overhead valve engine. But the price was £985

Pre-Rootes Singer. This 1949 SM 1500 is powered by a 1506cc overhead camshaft engine and cost £799

The world's first gas turbine car, the Rover JET 1 at Silverstone on March 9 1950. Maurice Wilks is at the wheel and Spencer Wilks (in black trilby) looks on. Frank Bell, the first turbine project engineer stands back to camera (extreme right)

the independent companies, whose share rapidly declined, tumbling from 11 per cent in 1946 to just over 4 per cent in 1955 and there it stayed. Rover and Jaguar headed the independents, both building on gains made in pre-war years. In view of their relatively small sizes both companies concentrated on long manufacturing runs to recover tooling costs. Therefore the Rover P4 range ran from 1949 until 1964 and, for instance, Jaguar's familiar 2.4–litre saloon shape, introduced in 1955, didn't cease production until 12 years later. Rover also had the advantage of having a second string to their manufacturing bow in the shape of the Land–Rover, introduced as a stop-gap measure in 1948, and selling strongly to this very day.

Sir William Lyons

But for all Spencer Wilks's success with Rover, the outstanding personality of these postwar years is Sir William Lyons (knighted 1956). Right from the birth of the Jaguar marque (though the original rendering was S.S.), in 1931, he offered the public utterly distinctive cars that he had styled himself, at a price they could afford. He also had a flair for choosing talented staff and as the marque evolved the cars performed as well as they looked. Before the war he had relied on Standard-based engines but the sensational bronze XK120 that graced Jaguar's stand at the 1948 Motor Show was of enormous significance because under the bonnet was the company's own potent twin-overhead-camshaft engine. Lyons's artistic sensibilities had demanded a

Sir William Lyons at his desk on the occasion of his knighthood in the New Year Honours, January 1956

power unit that also *looked* impressive, an attitude more of continental origin than British. The XK engine was a power unit in the traditions of the twin-cam Bugatti units of the '30s but, while the exquisite products from Molsheim were only built by the hundred the Jaguar company led Europe in the mass-production of twin-overhead-camshaft engines. Lyons was, above all, a manufacturer of saloon cars and it should not be forgotten that the XK engine was initially conceived to power the big Mark VII of 1951 at 100mph. The C and D type sports racers powered by this engine scored sensational wins at the Le Mans 24-hour races, reminiscent of the Bentley victories at the Sarthe circuit in the 1920s.

Sports car boom

The sports car market, enormously stimulated by the export trade, boomed in the '50s and '60s. We have already noted Standard's successful launch of their TR range and the market potential hadn't escaped Len Lord's attention. Therefore, he asked four companies to submit designs for a sports car employing BMC components. This was a familiar Lord tactic and Frazer Nash, Jensen, Healey and in-house MG were allotted the task but at the 1952 Motor Show he chose the Healey 100, the car becoming the *Austin* Healey overnight and a new BMC marque was born. MG's project was therefore sidelined but it finally emerged as the highly successful MGA of 1955.

Similar to the earlier T series of MGs was the good-looking Morgan sports

The MG TC was the spearhead of the sports car revival of the post war years. A delightful harvest time study of 1946

Below, MGs offering on a similar theme, EX 175, and also built in 1952. Under the bonnet was a 1250cc XPAG engine but three years later it blossomed as the MGA

The cars that the Austin Healey beat to the production line. *Left*, Frazer Nash's 1952 offering with 2660cc Austin engine though the steering column gear change was a jarring note. *Right*, the Jensen built car of the same year though it missed the Motor Show because some brake parts failed to arrive on time. A 1.2–litre A40 engine and MG gearbox were fitted. The photograph was taken in 1953 after a glass fibre hardtop was added

car from Malvern, but unlike Abingdon, Morgan did not possess the resources to tool up for all steel bodywork, as MG did with the A. Therefore, they continued building their cars by hand, the bodywork being traditionally shaped around an ash frame. Although marque interest faltered in the early '60s, the method of construction and the enduring styling of the '30s gradually emerged as an asset and, with an overall export emphasis shifted from America to the European market, demand soared.

Diametrically opposed to Morgan in concept was the alloy and glass-fibre-bodied Lotus. Whereas the Morgan marque represented a clear unflickering view of the pre-war years, the Lotus was an unashamed child of the '50s. Just as William Lyons and the Jaguar marque are inseparable, so Lotus and Colin Chapman are synonymous. Chapman, who left university in the late '40s with an engineering degree, began by building sports and sports racers in London's Hornsey in his spare time. The glass-fibre Elite coupé of 1957, superbly styled by accountant Peter Kirwan-Taylor, was the first road-going Lotus to be sold in quantity to the public, a position strengthened by the appearance of the Elan in 1963. The marque, along with Bristol, TVR and Reliant, are the only new makes born in the postwar years to still be intact at the time of writing (1980), a tribute to Chapman, who combines the rare qualities of a first-rate engineer with a businessman's cutting edge.

The first public appearance of the Healey 100 at the 1952 Motor Show. It received the Austin prefix for the occasion and a new marque was born

***Right*, of the 101,081 MGAs built no less than 91,547 were exported. Roadsters at London Docks await shipment**

***Left*, the Allard was another new postwar marque. In 1952 Sydney Allard (centre) won the Monte Carlo Rally in his P1 saloon. Rodney Walkerley, *The Motor*'s sports editor is on his right and co-driver Tom Lush to his left**

Colin Chapman at the wheel of a Mark 9 Lotus in 1955. He called his girlfriend Hazel 'my little Lotus blossom' and thus the cars . . .

If this had been the complete Lotus story it would have been remarkable enough, but the company also produced racing cars and spearheaded Britain's dominance with Cooper in the field of Grand Prix racing. Chapman graduated from Formula Junior single-seaters to Formula One cars and from 1960 he became a notable figure on the world Grand Prix stage. His cars were initially powered by Coventry Climax but when that company withdrew from the racing engine business, the legendary Cosworth DFV of 1967, financed by the Ford Motor Company, took over, contributing to Lotus's six Constructors' Championship wins.

Good times for the thoroughbreds

Without doubt the top-quality British sports car of the postwar years is Aston Martin. The marque had been in existence since 1922 and had experienced its fair share of financial problems. However, in 1947, David Brown purchased the company, taking it under the protective wing of his large gear-manufacturing and engineering combine and the outlook became considerably healthier. Soon afterwards he bought the Lagonda company, and its W. O. Bentley-designed, twin-overhead-camshaft, 6–cylinder engine, and a merger of the Aston Martin chassis and Lagonda power unit produced the DB2 of 1950, the first of a line of truly magnificent sports cars. The production of road-going cars and sports racers was the order of the day

Right, **the Aston Martin DB4, announced in 1958. Styling was by Touring of Milan and constructed at Newport Pagnell following the Italian company's Superleggera principles whereby a substructure of small diameter tubes was covered by a light alloy skin.**

Below, **Crewe products pictured on the occasion of the Society of Motor Manufacturers and Traders Jubilee parade through London in July 1946. Left to right, Rolls–Royce Silver Wraith sedanca-de-ville by H. J. Mulliner, limousine and Bentley Mark VIs**

H. HOBBS WILSON
BENTLEY
HTU955

and in 1959 Aston Martin won the World Sports Car Championship, the only British company ever to achieve this honour, a year that also saw the marque take the chequered flag at the Le Mans 24-hour race. Also, in 1959, came the completely new and fabled DB4. A magnificent pedigree had been maintained.

Quality was also the keynote of the Bristol, if not a sports car then a saloon with distinctive sporting qualities. When the Bristol Aeroplane Company decided to enter the car business, they wisely borrowed the best of German engineering by refining pre-war BMW designs, the 400 of 1947 clearly displaying its Teutonic antecedents. The cars were hand-made, in much the same way that Bristol had carefully built up their radial aero-engines. The BMW origins were gradually superseded with a rapidly-evolving body shape, though the ingenious 6–cylinder engine lasted until 1961 when it was replaced by a slightly modified Chrysler V8, and the original transverse-leaf independent front suspension was changed, simultaneously, to a more conventional coil-and-wishbone layout. Thankfully, this most individual of British makes is still with us.

Top of the quality ladder was Rolls–Royce which also embraced the Bentley marque. Production moved from Derby to Crewe after the war but, if these famous cars had been built in the manner employed in the '20s and '30s, it is unlikely that they would still be a major force in the land. Fortunately, a policy of rationalization laid down in pre-war years paid dividends and the Mark VI Bentley, with its pressed-steel body, was the first flowering of this philosophy. Clearly the future of the two makes depended on 'quantity' production and the Rolls–Royce Silver Cloud and Bentley S 1 of 1955 were a further step forward. (The radiator shells were the only outward differences between the two models.) The monocoque Rolls–Royce Silver Shadow and T Series Bentley of 1965 was yet a further recipe for survival in a difficult decade.

Opposite, **Alec Issigonis (right) at Cowley with the millionth Morris Minor in January 1961. The bonnet sides carry the 'Minor 1000000' legend. The car was finished in white**

Mini by Issigonis

At the opposite end of the automotive spectrum was the small, cheap, ingenious front-wheel drive Mini, launched by BMC in 1959. It proved to be the most revolutionary car in the history of the British motor industry and after 60 years of largely emulating the rest of the world's designs the Mini reversed this trend and has since influenced the course of small-car design across the globe. For more than 25 years front-wheel drive had been largely a continental preserve, but now it became a British one.

The driving force of the Mini's conception was, of course, Sir Leonard Lord. His interest in the production of front-wheel-drive cars dated back to 1948, at least, when, in his capacity as Austin's chief executive, he paid out £10,000 for a prototype front-wheel car called the Duncan Dragonfly. This highly professional small car was, for the most part, conceived by two engineers who had been engaged on Sir Roy Fedden's war-time car project which was abandoned in 1946. The Dragonfly was styled by Frank Hamblin while the mechanics were the work of Alan Lamburn. A BSA motorcycle

Right, the front wheel drive Duncan Dragonfly of 1948, purchased by Len Lord for £10,000. It was fitted with a transversely mounted two cylinder engine. Wheels are 12 inch and Alex Moulton was responsible for the suspension of rubber pastiles

engine was transversely mounted driving the diminutive 12-inch front wheels, while Alex Moulton was responsible for the car's rubber suspension. After Austin purchased the Dragonfly it was never again seen in public but it is interesting to note that Lamburn improved on his original concept in 1950, designing a new transversely-mounted power unit, embodying a gearbox contained within the sump, anticipating the Mini layout. Unfortunately this project never got beyond the drawing board although he sent the specifications to almost every car manufacturer in the world!

Above, **the Mini-Minor as it appeared in 1959. The Morris version is shown. Austin Minis had a waved pattern grille. The price? £496**

In 1956 Lord wooed the brilliant Alec Issigonis, designer of the Morris Minor, back from the Alvis company he had joined in 1952, to be BMC's chief engineer. The Suez crisis of 1956 triggered off petrol rationing at the end of that year which lasted until the following May. The result was that swarms of German bubble cars, with their economical aircooled engines, were soon seen in increasing numbers on Britain's roads. Lord therefore asked Issigonis to come up with a new small car. 'God damn these awful bubble cars' he told the designer, 'we must drive them off the streets by designing a *proper* miniature car.' At the same time Laurence Pomeroy, technical editor of *The Motor* was commissioned to come up with an alternative vehicle but the rear-engined miniature he conceived with David Hodkin of ERA was, he felt, in retrospect 'too ambitious'. Really intensive work on the Issigonis project began in March 1957 and, by October, two orange-coloured prototypes were on the road. In July 1958, one of these 'orange boxes' awaited Sir Leonard's inspection. 'What's this bloody contraption?' he asked its designer. Issigonis drove him round Longbridge . . . flat out. 'He shouted ''stop'' — got out — and said, ''make the bloody thing'' and walked away.' Issigonis was later to remember. The 'bloody thing', as Lord so characteristically put it, was, of course, the Mini. The totally individual offering bristled with ingenuity. A BMC A-series engine, Lord had insisted that a production unit be used, was mounted transversely across the car, power being transmitted to the 10-inch wheels via a gearbox positioned within the sump and sharing the engine's oil. The suspension, designed by Alex Moulton, was also unconventional and used rubber both in torsion and compression. The styling was uncompromisingly functional, which has perhaps contributed to the fact that the Mini is still in production. Issigonis's first car, the Morris Minor of 1948, was the first British model to sell a million, achieved in 1961, but it took the Mini only 6 years to reach this figure and to date, over 5 million have been made.

The Mini made its Motor Show debut in 1959, a year that also saw the appearance of the new Ford Anglia, with its distinctive inward-sloping rear window, and Standard–Triumph's Herald, an all-independent

Below, **the master stroke of the Mini's creation, the gearbox positioned within the engine's sump, thus drastically shortening the length of the combined units**

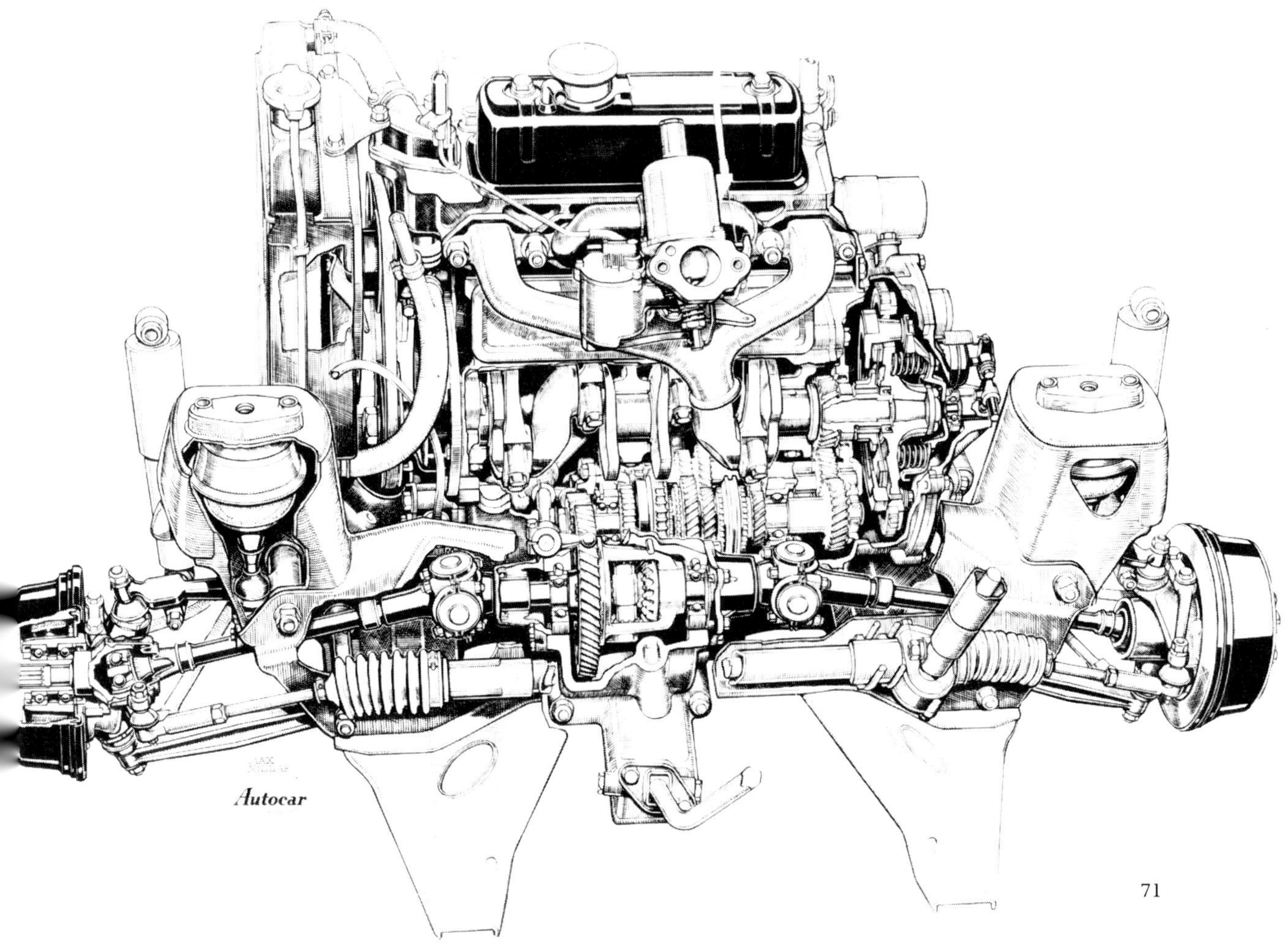

suspension saloon for £702. We left Alick Dick striving for independence in the body-building field and the Herald, it was noted, employed a chassis frame, instead of the more conventional monocoque. This meant that it was possible for the numerous body panels to be produced in different locations.

Len Lord, architect of BMC creation, pictured in the postwar years

The take-overs of the 1960s

But the '60s opened with a take-over that was to set the pattern for the decade. Daimler, with a bewildering range of models, had long been struggling for survival within the ailing BSA group and, in 1960, were taken over by the buoyant Jaguar company. Thus Britain's oldest marque name survives, albeit in badge-engineered form. 1960 brought a cash crisis for Standard–Triumph and Dick realized that unless the company could find a partner it would be forced out of business. He conducted talks with the American Motors Corporation, but, by the following year, Standard–Triumph had been engulfed by the ambitious Leyland Motors, who had been making commercial vehicles since 1897. In the '20s, Leyland had toyed with the car market and were concerned with such automotive extremes as the luxurious Leyland Eight and the utilitarian Trojan. In 1948 they had even contemplated taking an interest in the troubled Jowett company and there were talks with Rolls–Royce in 1959. Under Sir Henry Spurrier's direction, Leyland had done well in the commercial vehicle market in the postwar years and Standard–Triumph were soon to benefit from the take-over with a return to profitability in 1962.

American involvement within the industry was increased in 1964, with the Chrysler Corporation anxious to expand their European operations. In 1964, after securing a substantial shareholding in the French Simca company, they took a share in the ailing Rootes Group, which they increased in 1967 to give them effective total control. Rootes, who had had a damaging strike at their British Light Steel Pressings plant in 1961, also suffered when their new small car, the Imp, failed to find its form. It was a cruel irony that Rootes, who had never been regarded as technical innovators within the industry, should falter in the midst of producing an altogether new model embracing all-independent suspension and a rear-mounted overhead-camshaft aluminium Coventry Climax derived engine. The Imp staggered into production in the spring of 1963, but there were problems of reliability and it never sold in the numbers expected.

The early '60s were good years for BMC. They had followed up the Mini with the Farina-styled 1100 of 1962 and, by the end of the year, held a massive 44 per cent of the market. The 1100 was a further endorsement of the front-wheel-drive theme but with a new interconnected Hydrolastic suspension designed by Alex Moulton. The car looked right, was economical and possessed all the road-holding advantages of front-wheel drive and, even if it was criticised for its small boot, it mattered little. For the next 4 years it was to be Britain's top-selling car.

However, there were changes taking place in the Corporation's hierarchy. Sir Leonard Lord had been suffering from ill health for some years and, in

1956, his eventual successor, George Harriman, became deputy chairman and managing director, taking over the reins completely when Lord handed the chairmanship over to him in 1961. Sir Leonard, who became Lord Lambury in 1962, died 5 years later at 71 years of age.

Harriman, who had begun his career with Morris Engines at Coventry and then switched to Austin, was to guide the Corporation's fortunes in the '60s, along with Alec Issigonis, who was made technical director in 1961 and joined the BMC board in 1963. Issigonis, who had a high regard for Citroën front-wheel-drive technology, was convinced that, within a decade, pull instead of push would dominate and, though it took rather longer than he expected, it is now the order of the day.

The Corporation decided that, by adopting a policy of engineering excellence, they could present a noticeably different approach to Ford, their chief rivals at Dagenham. By contrast, the American company (and it became wholly so in 1961) offered conventional engineering and regular model changes. By producing technically more advanced cars that could be sold over much longer periods without major change, BMC calculated that the heavy re-tooling cost of developing new models could, accordingly, be reduced. But the profit margin on front-wheel drive cars was less than that on conventionally-driven ones. . . .

BMC were less lucky with their next model. Again a transverse-engine front-wheel-drive layout was chosen and the Austin 1800, announced in 1964, was a strong, roomy car, though perhaps the styling was an unhappy combination of the talents of Pininfarina and Alec Issigonis. When *Motor* magazine asked four members of the public for their first impressions of the car, two said that they didn't like the look of it, one going so far to describe it as 'an ugly beast'. But for whatever reasons the 1800 didn't match sales expectations. It should have sold 3500 a week but was only achieving about a 1000 by 1967. The jump from 1100cc to 1800cc meant that the massive gap in the BMC range was filled by the ageing and relatively pricy Farina-styled saloons. And into that slot, in 1962, moved Ford's new Cortina.

Dagenham were building on their gains of the '50s. The new 105E Anglia was selling strongly, though the Classic of 1961 was a less successful offering and about 126,000 examples, together with its Capri variant, had been made by the time production ceased in 1964. However, project Archbishop, as the new Cortina was coded, was an outstanding success. Its development was a meticulous operation, directed by Ford's product planning manager, Terry Beckett. A team of five was chosen to develop the Cortina package, which was a completely conventional front-engine rear-drive model, very much in the Ford mould. The clay model was approved in November 1960 and there were no significant hitches in development, except when production targets were increased by 50 per cent when the company realized that it had a winner on its hands. Job Number One, the first Cortina, came off the Dagenham production line in June 1962. The whole project had been conceived and executed within 19 months. The Cortina was a light-weight saloon, offered in either 2– or 4–door form, initially available with a 1200cc engine and later joined by a 1500cc unit. The bodywork, styled by Roy Brown, was nicely proportioned and rather racy . . . and the public loved it. In the first year 300,000 were sold, making the Cortina not only Britain's, but Europe's, fastest-selling car. Not surprisingly, 4 years later, in September 1966, the millionth Cortina was produced and that month the Anglia also achieved the same magical figure. A Mark II version of the Cortina, which appeared in 1966, was equally successful and, again, over one million were made during the following 4 years.

Sir Terence Beckett, who joined Ford as a graduate trainee in 1950, went on to become chairman of Ford of Britain

In 1966, therefore, American involvement in the industry consisted of Ford and Vauxhall, while Chrysler were progressively increasing their influence within the Rootes Group. They were ranged against BMC, who absorbed the Jaguar company, to create British Motor (Holdings) and a year earlier had taken over Pressed Steel, Britain's last independent manufacturer of mass-produced car bodies. The company, that had been set up in 1926 by the American Budd Corporation and William Morris at Cowley to provide body pressings for his group, began supplying the needs of the rest of the motor industry after Morris withdrew his interests in 1930. By 1935 Pressed Steel was wholly British owned. But this move by BMC caused some concern to the Rover company at Solihull. They were dependent on Pressed Steel for their bodywork, and perhaps they felt in a similar position to Standard, back in 1953, when BMC had taken over Fisher and Ludlow. The other growing force in the British-owned sector of the market was Leyland and when Rover received a take-over approach from them in 1966 it was accepted. Sir Donald Stokes, consolidating an enviable sales reputation, was already emerging as Leyland's driving force and from 1964 there were rumours of a merger between BMC, as they were then, and Leyland.

British Motor (Holdings) were faring badly. In 1966 their market share had dropped to 34 per cent and the following year it plummeted still further to 28 per cent, only just ahead of Ford. In 1967 the Cortina overtook the Austin–Morris 1100 as Britain's top-selling car, a year that also saw a complete integration of Ford's European activities.

The tragedy was that the Mini made little, if any, money for the Corporation. The complete story of its role within the Austin–Morris group has yet to be told, though Rob Golding in *Mini* reveals that it was making a loss around 1966 when it was thought to have been yielding a profit of about £15 per car. For, early in 1977, Colin Daniels, Leyland Cars finance director, was able to look at all the company's models on a world-wide basis. And, says Golding: 'He was able to argue without fear of contradiction that the Mini was losing money.' Although the car was profitable in Britain, exports were unprofitable and, therefore, the overall picture was one of deficit.

Perhaps it was a story of mini cars making mini profits and it is certainly true that the larger cars, which tend to produce larger financial surpluses, were relegated into second place by the Corporation in the face of the sophisticated, small, front-wheel-drive ones. To my mind it was far more the *total commitment* to front-wheel drive that was at fault. For, while the

Right, Ford's postwar expansion was directed by Sir Patrick Hennessy (left) seen here with Henry Ford II with a 105E Anglia, the 400,000th car produced at the new Paint, Trim and Assembly building at Dagenham

The Ford Cortina, launched in 1962 was the fastest selling car in the British motor industry's history. This Mark 1 car dates from 1965, the year that saw the Aeroflow heating and ventilation system introduced to the range

828 MOC

732 NJO

Above left, **getting it wrong, the Austin 1800 of 1964.** *Above*, **getting it right, the Farina styled A40 of 1958. 'The best looking product of the BMC' said** *Design* **magazine**

Left, Lord Nuffield, pictured in 1962, with the top selling Morris 1100. He died on August 22 of the following year, aged 85

Corporation was pre-occupied with pull instead of push, it ignored the traditionally-engineered but technically unexciting rear-wheel-drive cars. The later Leyland management were able to show this with the introduction of the conventional Marina of 1971. With the benefit of hindsight, that was the type of vehicle that should have been introduced in 1964 and not the 1800. Had front wheel drive been *eased* into production, the results might have been very different. The similarity with Citroën's collapse in 1935, following the introduction of the revolutionary *Traction Avant*, provides an inescapable parallel.

Yet another factor was the termination, in the '60s, of the Corporation's commitment to the Italian styling house of Pininfarina, which provided them with such shapes as the A40 of 1958 and the best-selling 1100 that derived from it. Design responsibility was, thereafter, centred at Longbridge and the outcome was such vehicles as the originally underpowered Maxi of 1969, that required some front-end surgery before it went into production, and the short-lived 3–litre saloon of 1968 that managed to sell a mere 10,000, when it, and the engine it shared with the equally brief MGC, was dropped after only 3 years' production.

All these factors, exacerbated by the scattered nature of the company and the labour problems that plague car makers throughout the world, contributed, I believe, to the group's subsequent take-over.

The inside story of the conception of the British Leyland Motor Corporation has been ably told by Graham Turner in *The Leyland Papers*. It

Triumph and tragedy. BMH's Sir George Harriman (left), chairman elect and Leyland Motors' Sir Donald Stokes, deputy chairman and managing director, announce the creation of the British Leyland Motor Corporation on January 17 1968. Stokes took over as chairman that September. It was the biggest merger in British industrial history

needs only to be said that there were a series of protracted negotiations, actively aided by Lord Kearton, chairman of the Industrial Reorganization Corporation. Eventually agreement was reached but this amounted to the virtual take over of BMH by Leyland.

When the announcement came in January 1968 it was described by Sir Donald Stokes and Sir George Harriman as a 'very happy merger'. In fact the negotiations had been protracted and, in their later stages, particularly bitter. But by sales value, British Leyland was the second largest motor manufacturer, after Volkswagen, outside the USA. The intention was to rationalize the range of vehicles and drop unprofitable models. But, above all, 'we are in business to make money', Stokes bluntly informed the press conference that broke the news of the new Corporation's creation.

We can now see that January 1968 represented the end of an era. The British motor industry has never been quite the same since. At Longbridge it was the end of the engineers' reign. From thereon the businessman took control while the rest of the world has followed the Mini's example, just as Alec Issigonis predicted. What was once the unconventional has now become the norm.

PART 2

A.C. 1908 to present date

The first A.C. was a passenger-carrying version of the Auto-Carrier, a provision orientated tricar that had been made since 1904. The firm's joint founders were John Portwine, a butcher, who supplied the finance, and John Weller, a talented engineer. The works of Autocars and Accessories Ltd. was established in London's West Norwood and, in 1907, the company's name was changed to Auto-Carriers Ltd. Thereafter, the tricar was offered with passenger accommodation, initially in tandem, but by 1909 the Sociable was available with side-by-side seating. In 1911 the company moved to Thames Ditton, Surrey and 2 years later their first 4–seater was announced, a Fivet-engined light car with gearbox mounted in the rear axle. After the First World War, in 1919, a 1.5–litre, side-valve Anzani engine replaced the original power unit and production continued until 1927.

But Weller's masterpiece, announced in 1919, was a new wet-liner, 6–cylinder overhead-camshaft aluminium engine. It was listed initially in 1.5– and 2–litre forms it reached production status in 1921, though only in the larger capacity. The same year S. F. Edge, who had spearheaded Napier's fortunes in pre-war days, became a director and took over as chairman and governing director in 1922. Weller and Portwine resigned and the company's name was changed to A.C. Cars Ltd. Edge embarked on an ambitious programme of racing and record-breaking and, in 1926, the Hon. Mrs Victor Bruce, with W. J. Brunell, scored Britain's first Monte Carlo Rally win in a 2–litre six A.C. The firm's name was changed, yet again, in 1927 to A.C. (Acedes) Cars Ltd., but 2 years later went into voluntary liquidation. Edge retired and, in 1930, two brothers, William and Charles Hurlock rescued the company, production re-starting in 1931. During the 1930s the Hurlocks offered a handsome range of sporting cars powered by the faithful Weller engine.

From the outbreak of the Second World War the company produced aircraft components for Fairey Aviation and, in 1940, a further factory was established on Taggs Island on the River Thames to manufacture guns, rocket launchers and radar equipment. After the war A.C. took a little time to get into their automotive stride and it wasn't until 1947 that their 2–litre saloon appeared. This was a well-equipped, though rather old-fashioned, cart-sprung model, which remained in production until 1956. By contrast, the Ace of 1954, Britain's first all-independently sprung sports car, was an up-to-the-minute offering. It was an effective revival of a pre-war name and A.C. sporting tradition.

The performance of the Tojeiro–Bristol sports racers had much impressed A.C.'s Derek Hurlock, Cliff Davis being the most successful exponent of the make during 1953. Vincent Davison, who worked for Tojeiro, had also produced an example, though his car was powered by a Lea-Francis

Left, **the most potent Ace, a Bristol engined example of 1959.** *Above*, **the Aceca, the coupé version, on its announcement at the 1954 Motor Show. Price was £1722**

engine, obtained from Connaughts. A.C.s therefore purchased this car, along with Davison, and proceeded to transform it into the road-going Ace. The power unit was hastily removed and a UMB series A.C. engine fitted in its place. This was the car that appeared at the 1953 Motor Show. Its handsome bodywork was inspired by the Touring-bodied Barchetta 166 'Inter' Ferrari. The Ace, with minor changes, went into production the following year and, later in 1954, the Aceca coupé version appeared. Bristol engines were available for both from 1956. In 1960 came a further variation, the long wheelbase 4–seater Greyhound coupé with coil-and-wishbone independent front suspension and A.C. or Bristol engined. When supplies of the latter unit were threatened the Ford Zephyr engine was offered in the Ace and Aceca.

In complete contrast to these sophisticated and potent products, A.C. made the 3–wheeler, B.S.A.-engined invalid monocar and in 1953 came the Petite, a 3–wheeler in the traditions of the old Sociable, but it never really caught on and production ceased in 1958.

Meanwhile the Ace was destined for another transplant. In September 1961, Texan racing driver Carroll Shelby, who had co-driven the winning Aston Martin at Le Mans in 1959, wrote to A.C. suggesting that the Ace would lend itself to V8 fitment. Later he became attracted to the 4.2–litre unit fitted to the Ford Fairlane, as new thinwall casting techniques meant that this iron engine weighed only 490lb with accessories. A.C.s responded positively and in the winter of 1961–1962 the prototype Cobra (Shelby had dreamed up the name) was built at Thames Ditton. Disc brakes were standardized with transmission, suspension and wheels being tailored to cope with the increased power. Demand was immediate and a production line for Cobras, though minus their engines and transmissions, was established. The A.C.s were then shipped to Shelby's works at Santa Fe Springs and later Venice, California for completion. The first 75 cars were fitted with 4.2–litre (260cu in) engines but for 1963 the model became the Cobra 289, a 4.7–litre engine being used.

This increased demand meant that A.C. dropped all their other models, the Weller-designed 2-litre six ceasing production 44 years after it was first announced! Meanwhile the Cobra gained rack and pinion steering and wider wheels and tyres. In October 1964 the Ace's original transverse leaf-spring-and-wishbone suspension was replaced by a

Top, **Bristol engine in situ. The front transverse leaf spring can be clearly seen.** *Above*, **driving compartment of a 1959 car**

new coil spring/wishbone layout from America. An even larger engine was fitted from October 1965, the Cobra 427 having a 7–litre V8 Ford, the title traditionally reflecting the engine's displacement in cubic inches. But the model was gradually becoming more Shelby and less A.C.

So for 1966 A.C. produced their own 427, an extended Cobra chassis being fitted with a handsome open 2–seater body styled and produced by Frua of Turin. The following year the engine was changed to one of 7016cc, the car becoming the 428, a fastback version being announced later in 1967.

A.C. production 1947–1967

2–litre		
2–door saloon		1142
4–door saloon		46
Drophead coupé		23
Experimental drophead coupé		1
Chassis only		25
Buckland tourer		59
	TOTAL	1296
Aceca, AC engine		152
Aceca, Bristol engine		169
Aceca, Ford engine		8
	TOTAL	329
Greyhound		*82
*includes round tubed prototype, flat six engined car and 2.6–litre Ford powered chassis		
Cobra		
Marks I and II		664
Mark III and chassis without bodies		350
Others		5
	TOTAL	1019
427 and 428		
Frua bodied (to 1973)		80

Ace 1954–1963

Specifications and developmental history

The A.C. Ace was announced in open 2–seater form. The chassis was formed by two 3–inch diameter tubes linked at the centre with a cross tube of similar diameter. Independent suspension — a transverse leaf-spring with lower wishbone layout, aided by telescopic shock absorbers — was employed front and rear. Girling hydraulic brakes were fitted and steering was by rack and pinion.

The Ace was powered by a 3–carburettored version of the famous 1991cc (65 × 100mm) 6–cylinder, ohc engine developing 85bhp. Leather-trimmed seats were provided and the car was priced at £1297. By the time that the car went into production the following year a few changes had been made. The headlights were raised and a forward-sloping radiator grille adopted. A Bishop cam steering box replaced the rack and pinion, while the body framing was strengthened and a little later the luggage capacity was increased. The 2–litre Bristol engine, developing 105bhp (also D type 120bhp option), became available early in 1956 and was listed later that year at £2011. In October 1957 disc brakes were offered as an optional extra on the front wheels for about £52 extra. From mid-1961 the 2.6–litre Ford Zephyr engine was available in five stages of tune ranging from 90 to 170bhp. It was priced at £1746. By the time the model ceased production in 1963 it was still being listed with three engine options: the Ford down to £1474, A.C.'s own power unit at £1565 with the Bristol-engined version costing £1873.

Check points

Body and chassis

The Ace's body is made from aluminium. It is, therefore, light but, as the suspension doesn't have much 'give', it can crack, which is usually at the shortest point between the door surround and rear wheel arch. You can also expect a certain amount of electrolytic action between the body and the steel chassis, particularly where it is wrapped around a member. The bodywork must have a symmetrical look, so stand well back from the car to get an overall picture. It's worth remembering that new shells are available, which is a major plus.

Simplicity is the keynote of the Ace's chassis because it is built up around two longitudinal tubes. However, one area where you might expect to find rust is where the exhaust pipe passes through or under the chassis outriggers. Check also the condition of the tube at the rear of the frame as it carries a hoop that supports the back half of the body.

Suspension, steering and brakes

The all-independent suspension is particularly prone to wear. This is because eight of its grease nipples have to be serviced every 500 miles and seven at 1000-mile intervals and, if this routine hasn't been rigorously adhered to, you're going to be in difficulties. It may be particularly noticeable at the back of the car causing rear-wheel steering. Fortunately, as production progressed the phosphor bronze bushes were replaced by Silentbloc ones. In some instances, bushes are secured by Allen keys and just tightening them up, if they have loosened off, can make a difference to the tautness of the suspension. By contrast, the shock absorbers give little trouble, except that the top mountings of the rear ones can fracture.

The Bishop cam steering isn't likely to present any major problems, though its mounting can loosen on the chassis.

Brakes shouldn't be troublesome and the only likely difficulty is with the Al-fin brake drums. On occasions, the steel linings can lift in their aluminium surrounds and, if faced with this particular problem, it is probably better to search around for second-hand ones rather than attempt a repair. On the disc-braked cars the rear wheel cylinders are much the same as those used on the short wheelbase Land-Rover.

Engine, gearbox and transmission

The Ace was fitted with three different engines during its production life, namely the Bristol, A.C. and Ford in that numerical order. The Bristol power unit and gearbox can be found on page 132 and the same remarks apply to the Ace. By far the best Bristol engine option is the 100D2 series, incidentally.

This leaves the A.C. and Ford, so let's first consider Weller's famous overhead-camshaft aluminium six. In view of its long production run it's a very reliable unit. Pre-war engines do suffer from corrosion around the wet-liners, but his shortcoming has yet to show itself on the postwar ones. Parts are in good supply, though crankshafts are now rather hard to obtain. A satisfactory oil pressure reading is 60psi at 3000rpm but be suspicious of any reading below 50psi. This engine was used in conjunction with a Moss gearbox, a robust unit that was also fitted in contemporary Jaguars. Over-drive was an option on all but the last examples and, if it isn't working in the Ace you're inspecting, the most likely fault is solenoid failure.

The Ford Zephyr engine was offered in five variations! Bottom of the range was the standard power unit with single Zenith carburettor, then came a stage-one iron head with triple S Us. Further variations were the aluminium Raymond Mays Conversion six-port high-compression cylinder head in three stages of tune with two or three S Us or triple Weber DCOEs. Those iron heads reworked by Tony Rudd could be had with a fluted aluminium rocker cover suitably inscribed *Rudd-speed*. Basic engine parts are, not surprisingly, cheap and plentiful, though you might have problems with non-standard spares. These engines were, on the whole, reliable.

You may find that a replacement engine has been fitted at some time in the vehicle's life, though this usually applies only in the case of Bristol-engined cars. However, it may help to mention the prefixes

The final engine variation was the 2.6-litre Ford Zephyr power unit. This is a 1962 Ace

to the engine numbers, which you'll find on the chassis plate, as they indicate the type of power unit originally fitted: A.C. engine, AE; Bristol engine, BE; Ford engine, RS (for Ruddspeed). An X suffix denotes left-hand drive.

The ENV rear axle is a sturdy and reliable component and should present you with few problems.

Spares

Fortunately, the A.C. factory will still service and even restore its older products and the Ace is no exception. A.C. engine spares are, therefore, available from this source. Bristol engine spares are still obtainable from Bristol Cars Services of London, while TT Workshops of Westbury, Wilts and Windley Restorations of Wokingham, Berks will undertake work on this rather specialist power unit and you'll find full details on page 132 in the Bristol section. Ford parts are considerably easier to come by, just one outlet for parts being Shepherds Grove Service Station of Stanton, Suffolk, details on page 143. Body parts are available from K. P. Autocraft of Weybridge, Surrey who have the original body jigs for the Ace. Also Classic Autos of King's Langley, Herts will undertake body repairs and replica work. The A.C. Owners' Club has naturally built up a large repository of knowledge relating to these cars, so if you own an Ace, it would be well worth joining.

Ace spares and services

A.C. Cars Ltd., High Street, Thames Ditton, Surrey (tel: 01-398 5621).
Classic Autos, 10 High Street, King's Langley, Herts. (tel: King's Langley 62994).
K. P. Autocraft, Brooklands Industrial Estate, Weybridge, Surrey.

A.C. Owners' Club, Membership Secretary, 4 Portsmouth Road, Camberley, Surrey.

Ace performance and consumption (Bristol engine), *The Autocar*, November 7, 1958.
Top speed: 117mph. Overall mpg: 21.6 for 1110 miles.

A.C. Ace production

A.C. engine	226
Bristol engine	463
Ford engine	38
TOTAL	727

Alvis 1920–1967

The first Alvis engine was designed by G. P. H. de Freville, who manufactured light alloy pistons under the Alvis trade mark during the First World War. These plans were bought by a 39-year-old Welshman, Thomas George John, who had founded the firm of T. G. John in Hertford Street, Coventry in March 1919. John had initially been a marine engineer, being works manager at Siddeley-Deasy of Coventry who manufactured Siddeley 'Puma' aero engines.

It wasn't until 1920 that the first Alvis appeared. This was designated the 10/30 and was a rather expensive 1460cc side-valver. The following year the firm's name changed to the Alvis Car and Engineering Company and a move was made to larger premises in Holyhead Road, Coventry. The bigger-bored 11/40 model was added for 1922 and, during the year, the Buckingham cyclecar was manufactured. But the really significant event of this 12-month period was the appointment of Captain G. T. Smith-Clarke as chief engineer and W. M. Dunn as chief designer. They immediately started work to re-design the engine; the outcome was the famous 12/50 model of 1923. This was powered by a 1496cc four-cylinder, overhead-valve engine, an example winning the 200 Mile Race at Brooklands in October 1923 at 93.29mph. The model certainly got off to a flying start but the same couldn't be said for the firm's finances and a receiver was appointed in July 1924. Alvis, like Rolls–Royce and Bentley, only built their cars in chassis form, the customer dictating his preference for a particular body style or design. A debt of £5000 owed to coachbuilders Cross and Ellis had precipitated the crisis though total unsecured debts amounted to £219,000. A scheme of reconstruction was approved and the years between 1926 and 1929 were ones of steady company growth.

In 1925 Alvis produced a technically-advanced low-slung, front-wheel-drive sprint car and the following year a 1.5–litre Grand Prix car, with unusual horizontal valves, was produced. The outcome of these experiments was the Front-Wheel-Drive model, which was the first British car to offer this facility. It was powered by a specially-designed 1.5–litre, overhead-camshaft engine that could be obtained with or without supercharger which was also specially tailored. However, the car was not a commerical success and production ceased in 1930.

A 6–cylinder Alvis appeared in 1927 and the Silver Eagle of 1929 was the forerunner of the Speed 20, Speed 25, Crested Eagle and 4.3 litre models of the '30s. These finely engineered sixes were always in the forefront of technical innovation, and in 1933 featured independent front suspension and an all synchromesh gearbox: two British firsts.

While the early '30s were fairly profitable for Alvis (in 1934 a record number of 1110 cars left the factory), John realized that if the company was to remain on a sound footing, diversification was essential. Therefore, in 1935, the firm's issued capital was increased to £370,000 so that Alvis could enter the aero-engine business and a new factory was specially built for this purpose. Although an Alvis-designed engine was produced it was not taken up by the government, while a 9–cylinder radial, the Leonides, did not come into its own until after the war.

This change in manufacturing emphasis was mirrored by the company's change of name. In 1936 the less automotively-committed Alvis Ltd., replaced the Alvis Car and Engineering Company. Although large losses were sustained in building up the aero-engine business, the firm moved into the black in 1939. These years saw Alvis enter the military vehicle market, yet another diversification that would, in the long term, prove highly profitable.

Not that we should lose sight of developments on the car front. George Lanchester joined the company and he, assisted by Harry Mundy, were largely responsible for the Silver Crest and the 1842cc 4–cylinder 12/70 which appeared for the 1938 season.

Alvis's Holyhead Road Factory was badly damaged by bombing in November 1940. However, the aero-engine factory produced Rolls–Royce Kestrel, Merlin and Griffon engines and was also responsible for a chain of factories working on their repair and overhaul. The company that emerged from the war was very different from the one that entered it. It was now geared to the production of heavy vehicles and aero-engines as well as cars. Aero-engine production soon got under way with Alvis's own Leonides radial proving a firm favourite with the aircraft industry.

Alvis re-started car production with just one model, the TA14 of 1946. This was a 4–door saloon with bodywork by Mulliners of Birmingham and powered by the pre-war 12/70 engine, though increased slightly in capacity to 1892cc. Rather surprisingly it was cart-sprung all round. In 1948 drophead coupé versions by Tickford and Carbodies appeared while 1949 witnessed the appearance of the TB14 2–seater sports car with twin carburettored engine. It was hardly in the Alvis traditions, with a grille in place of the traditional radiator shell, and was very much geared for export.

A new model, which superficially resembled the old Fourteen, appeared at the Geneva Motor Show in 1950. This was the TA21 3–litre, again available in saloon form by Mulliners with a drophead by Tickford. The box-section chassis was new, with no less than six cross-members being used, while coil-and-wishbone independent front suspension was employed. Lockheed hydraulic brakes were fitted and Burman Douglas worm-and-nut steering used. The engine was also new being a 2993cc (84 × 90mm) overhead-valve six, with twin Solex carburettors, developing 86bhp. Inside, leather seats and walnut dashboard were in the best Alvis traditions. Price was £1597.

In 1952 the twin Solexes were replaced by twin SUs and later the compression ratio was raised from 6.9:1 to 8:1. In 1952 a Roadster version, similar to the earlier TB14, was introduced but with a traditional radiator shell in place of the grille. This mostly export model was designated the TB21. In February 1953 came the TC21 saloon with a compression ratio of 7:1 and twin carburettors, though many TA21s were modified to these specifications. At that year's Motor Show a 100-mile-an-hour version, the TC21/100, named the Grey Lady, was introduced and had an 8:1 compression ratio, higher rear-axle ratio and distinguishable by its slimmer door pillars, bonnet scoops and Rudge Whitworth wire wheels. Production of the model ceased in October 1954 after Mulliners of Birmingham, who built the saloons, concluded an agreement with Standard–Triumph earlier in June, which gave the car company exclusive rights for their entire output of bodies.

Meanwhile Alvis were forging ahead with their other activities. Profits stood at £141,232 in 1952 with Saracen armoured car and aero-engine production dominating the manufacturing scene. The fact that car output accounted for a minute proportion of the company's profits can be gauged by the fact that in 1954–1955 they amounted to

The Graber styled 1959 TD21 built by Park Ward and costing £2993

over £150,000, though practically no cars were made over this period.

With Mulliners confining its manufacturing energies exclusively for Standard–Triumph and David Brown having taken over Tickford (who built drophead coupés for Alvis) in 1955, the problem was clearly one of body supply. Fortunately, since the early '50s the Swiss coachbuilders Herman Graber of Berne had been producing a few of their individually-bodied cars for continental customers; drophead coupés, 2– and 4–door saloons being offered. An agreement was therefore reached between Graber and Alvis the manufacturing rights being assigned to the car company and jigs, tools and drawings were purchased. Arrangements were put in hand for Willowbrook of Loughborough, who built motor coach bodies, to produce 2–door saloons to the Graber design, but with a £3500 price tag there were few takers for this TC/108G.

Meanwhile, behind the scenes, Alec Issigonis was designing a new Alvis saloon, his stay with the company spanning the years 1952–1956. Designed to carry five passengers in comfort, the car was conceived with a 1750cc 4–cylinder engine or 3.5–litre overhead-camshaft V8, the prototype being powered by the latter. The transmission was unconventional with a 2–speed gearbox mounted in the rear axle and a Laycock over-drive unit operating on both speeds. The gear 'lever' was situated on the driver's right. Suspension was initially rubber, produced under Moulton patents, though later there was interconnection between the front and rear units — the first steps towards a Hydrolastic system which saw fruition on the BMC 1100 of 1962. The project floundered when Alvis discovered that the cost of a new factory for its production was beyond them and Issigonis returned to BMC at Sir Leonard Lord's behest. And we all know what happened then. . . .

But back to production cars. Willowbrook continued to build the Graber saloon for 3 years but output really got underway after October 1958 when Park Ward took over the body contract. The car's price was reduced by £500, seating and visibility were improved, the chassis strengthened and the engine was fitted with a larger oil pump. A gearbox that was shared with the big Austin Healey was introduced, hydraulic clutch operation appeared and Lockheed disc brakes were fitted at the front. The handbrake was transferred to a position between the front seats; hitherto the pistol-grip type had been employed. Designated the TD21, the model was available as a saloon and drophead coupé. In March 1959 changes to the model were announced, a new cylinder head with six inlet ports was standardized and fitted with twin thermostatically-controlled SU carburettors: 120bhp was claimed.

Profits for the year ending July 1960 were the highest ever at £267,488, the Leonides aero-engine seeing service all over the world and one was also used to power the first Hovercraft. In 1961 Alvis signed an agreement with Harry Ferguson Research Ltd., for the manufacturing rights of his 4–wheel drive system and the use of the Dunlop Maxaret anti-locking device.

In August 1962 came the Series II TD21 model with Dunlop disc brakes all round, while the fog lamps were recessed into the air intakes and the rear lamp layout re-arranged. Under the surface, changes were made to the body construction, the framework and panelling being aluminium instead of the more traditional wood and steel. A ZF 5–speed gearbox was added for the 1963 models.

At the following year's Motor Show, a Series III version of the 3–litre appeared. This was the TE21 with distinctive, vertically-mounted twin headlamps, echoing the Graber-bodied cars exhibited at Geneva earlier in the year. Improvements were made to the steering and front suspension, and the engine's power was boosted to 130bhp.

Beautiful though they were, Alvis cars were gradually running out of time and their fate was sealed with Rover's purchase of the company in July 1965. The approach came from Alvis and negotiations were conducted swiftly and efficiently. This did not prevent the TF Series IV appearing in March 1966 having re-positioned controls and instruments and a larger heater. Power output was now listed at 150bhp, the engine's compression ratio being raised to 9:1 and an electric fan replaced the conventional one. ZF power-steering was also fitted. But the end was in sight and the last TF21 left the works on September 29 1967.

So that was the end, or was it? In 1966 Rover produced a replacement for the ageing 3–litre Alvis. Styling director David Bache took a Rover 3500 (P6B) underframe and mechanics and designed a good-looking two-door coupé, complete with hatchback. The Alvis GTS was a tragic 'might have been' as the Leyland takeover of Rover in 1967 came just at the wrong time for the car, which for some reason was known as Gladys! She didn't fit into the re-drawn plan, though the one and only example does survive as a reminder of what was to have been the next generation of Alvis cars.

The Series IV TF21, announced at the 1966 Geneva Motor Show. There were interior and suspension improvements and a 150bhp engine

Alvis production 1946–1955

TA14	3210
TB14	100
TA21	1310
TB21	30
TC21	250
TC21/100	475

TC/108/G 1956–1958
TD21 1958–1963
TE21 1963–1966
TF21 1966–1967

(For specifications and developmental history see previous text)

Check points

Body and chassis

First is the headlamp area, checking particularly in a line from the outer part of the rim down to the bumper. Then examine the base of the front wing and continue the inspection along the bottom of the sill (a terrible rust spot) to the rear wing. The doors of the Series I TD21 were timber-framed and used steel panels and are, therefore, far more susceptible to rust than later models employing aluminium panels. The rust really attacks the Alvis at the rear and occurs particularly where the inner rear wing section curves over the wheels to join the outer wing. In other words the perils of double skinning. Consequently the metal tends to rot outwards in a radius from the back wheel arch to a depth of about 3 inches. The first indication that trouble is on the way is a small series of indentations around the arch, caused by corrosion forming around the spot welding holding the inner and outer wings together. A vital boot check should not be overlooked. Examine the floor for rust breaking out in the corners and the sparewheel container is another particular trap for corrosion. Also examine the rear valance behind the rear bumpers.

Fortunately the chassis is extremely robust and

Top, **underbonnet view of the TD21 3–litre engine.** *Above*, **the beautifully appointed interior. 2-speed windscreen wipers and 'screen wash' were standard fitments**

isn't liable to present any problems, except where it sweeps above the rear axle. Holes can appear in the vicinity of the bump-stop rubbers.

Suspension, steering and brakes

The independent front suspension is expensive to overhaul and the most likely area for trouble is the lower wishbone joint where the kingpin swivels. Bad examples have been known to display a quarter of an inch of slack. The top wishbone joint can also suffer to a lesser degree, while the kingpin itself also wears. Another problem is that the front coil springs are inclined to tire, often caused by the tapered packing which can be introduced to compensate for loss of rebound. Unfortunately this can upset the spring rate. Fitting a new spring is something of a major exercise as the bottom wishbone has to be removed.

The steering box wears well and wear can be taken up in it and the idler. However, if slackness does develop the problem is more likely to occur in the idler through a worn ball race.

Brakes are reliable and don't present any special problems.

Engine, gearbox and transmission

The engine isn't known for any particular shortcomings but it's worth mentioning that the first few TD21s had the same engine as the previous TC21 which wasn't fitted with the full flow oil system of later examples. It is also possible that some TDs have been fitted with the TE21 cylinder head with larger valves and better breathing. No less than 15bhp extra was claimed for these improvements. Oil pressure should be about 40psi at 2000rpm.

Gearboxes haven't any particular vices though parts for the later ZF 5–speed boxes are getting expensive. Rear axle problems are rare.

Spares

When Alvis ceased car production, Red Triangle Auto Services took over official servicing and as a parts agency and they can supply practically all wearing parts on the TD series, whilst also catering for other models. They maintain close contacts with the Alvis Owner Club and can, therefore, remanufacture parts when demand reaches sufficient proportions. Hanns of Bridport, Dorset also specialize in postwar Alvis spares and restorations.

Alvis spares and services

Hanns of Bridport, Foundry Lane, Bridport, Dorset DT6 3RP. (tel: Bridport 22601).
Red Triangle Auto Services Ltd., Common Lane Trading Estate, Kenilworth, Warwickshire (tel: Kenilworth 57303).

Alvis Owner Club. M. J. Cummings, The Hill House, Rushock, near Droitwich, Worcestershire.

TD21 performance and consumption, *The Autocar*, October 16, 1959.
Top speed: 103.5mph. Overall mpg: 18.3 for 1044 miles.

Alvis TC/108/G, TD21, TE21 and TF21 production

TC/108/G		30
TD21	1958–1963	1060
TE21	1963–1966	350
TF21	1966–1967	105

Armstrong Siddeley 1919–1960

Armstrong Siddeley built their first car, a massive 30HP machine, in 1919. The marque had been created through a merger of Armstrong Whitworth's automobile interests of 1906–1915 with the Coventry firm of Siddeley–Deasy, production being centred at the latter's Parkside works.

John Davenport Siddeley (1866–1953) was educated at Altrincham County High and Beaumaris and joined the Humber Cycle Company as a draughtsman in 1892, moving to the Rover company 4 years later. He conducted a Daimler in the famous 1900 Thousand Mile Trial and in 1902 he began selling cars under his own name, though in reality they were thinly disguised Peugeots. They were later produced by Wolseley and in 1905 he succeeded Herbert Austin as that company's general manager. Early in 1909 he left Wolseley and with Captain H. H. P. Deasy, the English importer of Swiss Martini cars, formed Siddeley–Deasy Motors in Coventry. Deasy also marketed the Deasy car from 1906, designed by E. W. Lewis, late of the Rover company. The resultant cars were known as Siddeley–Deasys from 1912.

During the First World War the company took up aero-engine production, the Puma being the first to bear the Siddeley name. The Newcastle-upon-Tyne firm of Armstrong Whitworth (itself the product of amalgamation in 1897) supplied cylinder-head castings and later fully-machined units for the Puma and Siddeley was much impressed by the quality of the work. Consequently, after the war the companies decided to link up to produce aeroplanes, aero-engines and motor cars. Armstrong Whitworth Development Co. Ltd., was created in 1919 and its subsidiary was Armstrong Siddeley Motors who took over where Siddeley–Deasy had left off.

Of all these projects, cars were Siddeley's main interest, followed by aero-engines, while the Armstrong Whitworth aircraft were looked upon as a useful outlet for the company's radial power units. In 1920 Whitley Aerodrome south of Coventry was purchased and aircraft production was moved there in 1923.

Car output started in 1919 with the aforementioned 30HP overhead-valve six, while a scaled-down 2318cc model of 18HP appeared for 1922. This was subsequently uprated to 2870cc in 1926, becoming the 20HP model 2 years later. These sixes were supplemented by an 1852cc 4–cylinder 14HP car for 1924, and it remained in production until 1929. A side-valve 15HP six joined the range in 1928 and the following year the similar valved 12 of 1236cc appeared. This latter model, which remained in production until 1936, was to prove the company's most popular model of the inter-war years.

In 1929 the Wilson pre-selector gearbox was offered as an option on the Armstrong Siddeley, the fitment being standardized in 1933. This was historically appropriate as Armstrong Whitworth had produced the Wilson-Pilcher car of 1904–1907 with its epicyclic gearing and pre-selection, which was also the product of the fertile mind of Major W. G. Wilson.

The original 30HP model was replaced for 1933 by the interesting Siddeley Special which used a 4960cc overhead-valve 6–cylinder Hiduminium engine. This aluminium alloy had been developed by Rolls–Royce for aero-engine production. Only 253 examples of the Special were produced before the model ceased production in 1938. The 20HP model of 1928 had been uprated to 3190cc in 1932 and this remained available until 1936. It was replaced by the 20/25 with 3670cc power unit; output continued until 1939. Meanwhile the 12 had been replaced in 1937 by a 1666cc overhead-valve 6–cylinder 14HP model. Then, for 1939, came the 16HP with a similar body style to the 14 but with a 1991cc 6–cylinder engine. The 2780cc 20 appeared at the same time (the 20/25 had been known as the 25 since 1937). These models, together with a 2394cc 17HP car of long standing, rounded off the Armstrong Siddeley range at the outbreak of the Second World War.

Throughout the inter-war years aero-engine and aircraft production grew apace and in 1927 Siddeley moved firmly into control, the company's name being changed from Armstrong Whitworth Development to Armstrong Siddeley Development. In 1928 he acquired another aircraft manufacturer, A. V. Roe, at the time a subsidiary of Crossley Motors. The same year John Siddeley founded another important company, High Duty Alloys, with Wallace Devereux as chairman. Devereux had previously been works' manager at the old established firm of Peter Hooker, who Siddeley had, on occasions, bailed out when they found themselves in financial difficulty. High Duty

Alloys later took up the production of the Rolls–Royce-developed aluminium alloy, Hiduminium, supplying it to the aircraft and later the motor industry. Its use in the Siddeley Special's engine now becomes more logical. The culmination of Sir John's work (he had been knighted in 1932) came in 1935 with a merger with Hawker Aircraft and the creation of the Hawker Siddeley group. He finally retired the following year and in 1937 was created Lord Kenilworth.

During the Second World War the group produced some of the famous aircraft of the conflict: the Hurricane figher, the Whitley bomber and the Lancaster that replaced it. Therefore, it was not inappropriate that, the week the war in Europe ended, Armstrong Siddeley announced their new models bearing names that echoed those of their war-time aircraft. The 16HP range embraced the Lancaster 4–seater saloon (by Mulliners of Birmingham), the Hurricane drophead coupé and Typhoon sports saloon. These were mounted on the same chassis with torsion-bar independent front suspension and with the engine bored out to 2309cc, the model becoming 18HP for 1950. Then the Typhoon was discontinued in 1950, and the Lancaster was replaced in 1952 by the Whitley razor-edge saloon, which remained in production until 1954.

Behind the scenes W. O. Bentley had been commissioned to design a replacement. This was, in effect, a 3–litre (84 × 90mm) version of his dohc, 6–cylinder, postwar Lagonda engine mounted in a similar chassis though later a de Dion rear axle appeared. But it never materialized and the model announced at the 1952 Motor Show was one of the company's own design.

The Sapphire employed a 3443cc 6–cylinder engine with inclined valves mounted in a hemispherical cylinder head and operated by pushrod from a single high-mounted camshaft. Two gearbox options, manual or pre-selector were offered; the pre-selector unit was electrically operated. A separate chassis was still employed and wishbone-and-coil spring independent front suspension featured. Initially available as a 4–door saloon, it was joined by a limousine version in 1955. Power-assisted steering arrived the following year as an option, a notable British first. By the time that the Sapphire ceased production in 1958, it proved to have been the company's best-selling model.

Dignity from Coventry. A 1950 Armstrong Siddeley Hurricane 18hp drophead coupé. Price was £1246

The range was extended for 1956 with the arrival of the Sapphire 234 and 236. These were based on a common chassis and body with coachwork in Hiduminium with steel wings. The 234 used a 2290cc 4–cylinder version of the existing six, while that used in the 236 was a minor development of the earlier 18HP model. Sales were disappointing though and the four and the six were phased out in 1958 and 1957 respectively.

Last of the line was the disc braked Star Sapphire available as a saloon and limousine and produced in 1959 and 1960. The body style evolved from the 346, as the Sapphire had been known since the arrival of the two smaller models, for 1956. Power was provided by a 4–litre version of the pushrod six.

A tailing off in demand caused the factory to slow production. Then came the creation of Bristol Siddeley Engines in April 1959, formed by merging Armstrong Siddeley Motors with Bristol Aero Engines. (In the interests of historical continuity it is worth recording that, in January 1968, the Hawker Siddeley Group sold their share in Bristol Siddeley to Rolls–Royce, who had already acquired the Bristol Aeroplane Company.)

The last Star Sapphire saloon chassis was completed on July 5 1960, and the final limousine frame on September 6. After 41 years Armstrong Siddeley was no more.

***Above*, the Lancaster saloon, the most popular version of the 16/18hp theme. *Right*, a 1954 Sapphire six light saloon. A four light car was also available**

Armstrong Siddeley production 1945–1960

16 and 18HP models

Hurricane		2450
Lancaster		3472
Typhoon		1703
Whitley		2609
Station coupé		958
Utility coupé		708
Limousine		125
Shooting brake		1
Unidentified		1
Chassis only		112
	TOTAL	12,139

Sapphire 234	803
Sapphire 236	609
Star Sapphire	902

Sapphire 1953–1959

Specifications and developmental history

The Armstrong Siddeley Sapphire was announced at the 1952 Motor Show as a 4–door 5/6–seater saloon mounted on a box section chassis. Independent front suspension was by coil-spring and trailing wishbone and Girling hydraulic drum brakes were fitted, as was Burman re-circulating ball steering. The engine was a 6–cylinder unit of 3435cc (90 × 90mm) with inclined overhead valves operated from a side camshaft, developing 120bhp. Two transmission options were available: an electrically-operated pre-selector gearbox and an all-synchromesh unit. The interior was well finished in leather with wooden door fillets and dashboard. Price was £1728. In September the following year a twin-carburettored version was announced and compression ratio was increased from 6.5 to 7:1, producing 150bhp. A six light saloon was also offered. For 1955 a Mark II version, from chassis number 343750, with an automatic transmission option was available. Changes included a re-styled instrument panel, 12-inch diameter brakes (in place of 11-inch ones) with Clayton Dewandre servo assistance, improved carburettor linkage and Metalastik bushes in the front suspension system. In March 1955 a 7–seater limousine version of the Sapphire was announced, with the wheelbase lengthened from 9ft 6in to 11ft 1in. Automatic transmission was standardized. Models for 1956 were offered with Girling power steering as an option. At extra cost were electrically-operated windows and Telaflo high-duty shock absorbers. Two new models were introduced for 1956 (the 236 and 234) so the Sapphire became known as the 346 for that year, and until it ceased production in 1959.

Check points

Body and chassis

Sapphires do rust, but the problem can look worse than it really is! One big advantage is that the car is, for the most part, bolted together, rather than welded, this applying to the wings, door pillars and sills. Perhaps the most serious areas to rust are the main sills. Their role is a particularly important one as they support the body structure. You can see the sills' outer face when the doors are open. Another of their functions is to maintain the correct door apertures. A reliable indication of bad rusting is when the front door catches on the windscreen pillar, though make sure that the trouble isn't being caused by a broken hinge before checking for suspect sills. The really weak points are the inner faces as lightening holes allow water and road grit to get inside the sills, blocking the drain holes and preventing sludge and similar silt from draining away. Consequently, the bottom tends to rust out completely. In bad cases the corrosion can also weaken the base of the main door pillar. A word of warning though. If you do buy the Sapphire you're inspecting, and the sills require replacement, *remove only one side at a time otherwise the body could twist or sag beyond repair.*

As far as the wings are concerned, the front ones should be checked along their lower edges, adjoining the front doors, as rust is caused by mud thrown back from the wheels. The bottom of the doors can give trouble and the front of the rear wings should also be checked. Also don't forget to examine the edges of the boot floor as they too rust. Another vulnerable point is the heater box. Water gets in and you know you're in trouble as your feet get wet when it rains! The Sapphire's chassis is remarkably robust though careful inspection of the rear spring hangers and surrounding metal is important.

Suspension, steering and brakes

The model's independent front suspension is prone to king-pin wear and, though these are available, they're not cheap. Lubrication should be with oil rather than grease, incidentally. All the other bushes are renewable.

The steering box is fairly trouble free and, if slackness is experienced, then it is more likely to be the idler. A rather more unusual symptom such as front wheel imbalance, incorrect tracking or worn wheel bearings can be caused by disintegration of the Metalastik front engine mounting. This causes excessive engine vibration, which upsets the steering balance.

Brakes should give little trouble, though the Clayton Dewandre vacuum servo fitted to the Mark II cars can be troublesome. It is worth recording that a few cars were fitted with Girling front disc brakes.

Engine, gearbox and transmission

The engine is a reliable unit, but the timing chain is noisy and difficult to quieten. A water leak occurs sometimes between the inlet manifold and cylinder head as the rubber gasket sandwiched between the two can perish with age. An acceptable oil pressure reading is 30–40psi at 30mph, the Mark I's gauge being graduated 0–100 and the Mark II, 0–60.

Transmission can be either pre-selector, manual or later on the Mark II, automatic. Initially the pre-selector gearbox was by far the most popular option. However, the automatic became increasingly favoured on the Mark II cars. Let's take the pre-selector first. The box should be quiet in operation and remember that the 'clutch' pedal travels back varying distances in each gear. Loss of all gears is usually caused by blown fuses or bad contacts on the fuse itself and a stored vehicle is particularly susceptible to this. The fuse box, is located on the bulkhead under the bonnet, and the contacts should be cleaned. If any fuses are blown, then they must be replaced. Fuses are rated at 25 amps and the contacts should be cleaned every 12 months. Alternatively, the gearbox solenoids could be faulty or stuck. The switch on the steering column can also be troublesome and usually the fault can be traced to greasy contacts.

The manual gearbox is normally well-behaved though the lever, mounted on the steering column, does suffer from wear in the linkage. Similarly, the Rolls–Royce type automatic gearbox gives little trouble. It shouldn't jerk into second gear and the fluid flywheel should be smooth and not judder. Rear axles give reliable service.

Spares

The Armstrong Siddeley Owners' Club was fortunately able to purchase the entire stock of car spares from Rolls–Royce, their eventual owners, which is a major advantage. The club tries to keep all parts in stock, with drawings of individual components being recorded on microfilm. Manuals and workshop lists are available for all postwar models.

Armstrong Siddeley Owners' Club Ltd., M. J. Hubbuck, 90 Alumhurst Road, Bournemouth, Dorset (tel: Bournemouth 763413).

Sapphire performance and consumption, *The Autocar*, July 17, 1953.
Top speed: 91.25mph. Overall mpg: 18 for 376 miles.

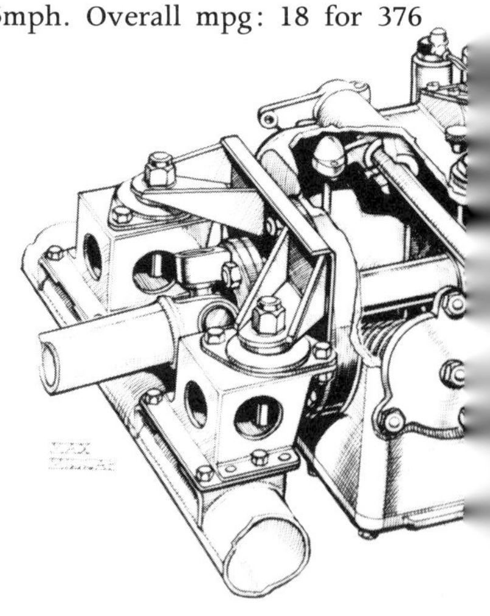

Armstrong Siddeley Sapphire production

Six light saloon	6869
Four light saloon	377
Prototype short chassis limousine	1
Long chassis limousine	381
Oil company special pickup	45
Chassis supplied to specialist coachbuilders	
Ambulances	3
Hearses	14
Long chassis pickup	1
Hooper body	1
Experimental and prototypes	5
TOTAL	7697

Right, **driving compartment of a 1952 four light Sapphire.**

The model's distinctive 3.4–litre six cylinder engine. The camshaft operates both inlet and exhaust valves. Gear selection in this pre-selector gearbox is made by five solenoids

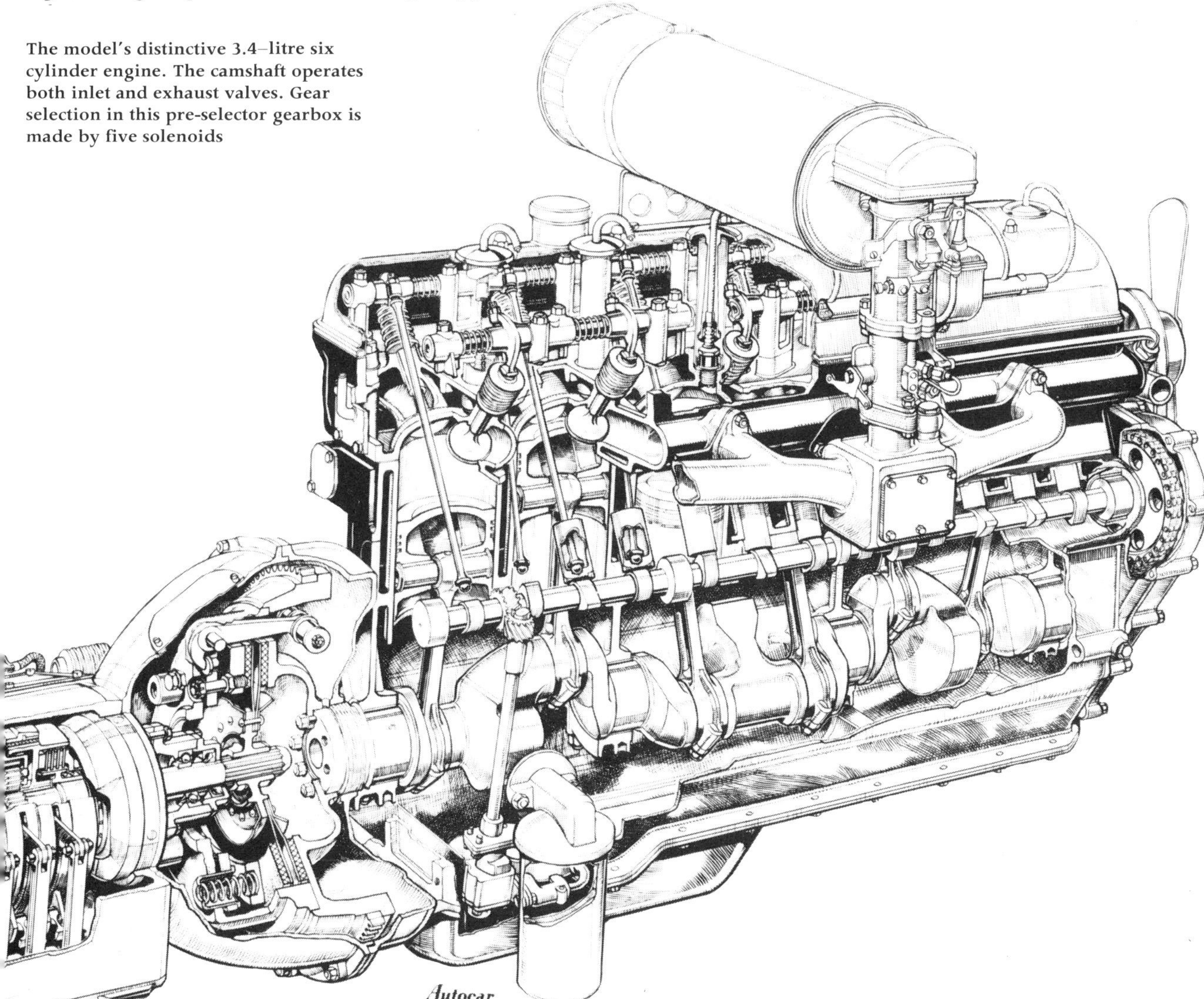

Aston Martin 1922 to present day

The original Aston Martin was built in 1914 by Lionel Martin, the prefix echoing its creator's successes in the Aston Clinton hillclimb. This first car used an Isotta-Fraschini voiturette chassis, powered by a 1.4–litre Coventry–Simplex engine. Martin was a partner in the firm of Bamford and Martin Ltd., (formed in 1913) who held a Singer agency in London's South Kensington. However, production didn't begin in earnest until 1922, the company's first offerings being mainly racing and experimental machines. The first cars offered for public sale had 1.5–litre side-valve engines though more exciting were the twin-overhead-camshaft cars built for Count Zborowski, which competed in the 1922 French GP. Single-cam variants were also available. These activities absorbed much money and in 1924 the original company was purchased for £10,000 by Lord Charnwood for his son, the Hon. John Benson, who took over a design role and explored further variations of the twin-cam theme. Unfortunately, finances ran out again and a receiver was appointed after the firm's appearance at the 1925 Motor Show. Propositions were received from Vauxhall Motors, the Bristol Aeroplane Co., and even as far afield as Donnet et Zedel in France. The moribund company was eventually purchased by William Somerville Renwick for £3600 and in 1926, the firm's name was changed to Aston Martin Motors Ltd. Renwick had previously met up with Augustus Cesare Bertelli at Armstrong Siddeleys and the two engineers formed their own company, Renwick and Bertelli Ltd., at Tyseley on the outskirts of Birmingham. Here they had built just one car, the R and B, which was nicknamed the Buzz Box. Under the bonnet was a 1.5–litre, single, overhead-camshaft engine designed by the two partners. This R and B engine formed the basis of all production Aston Martin's during the remaining inter-war years.

Benson, who stayed on after the reconstruction, soon clashed with Renwick and they both left the company (by then established at Feltham, Middlesex) leaving Bertelli solely responsible for design. Finances were always unstable and the company was re-formed as Aston Martin Ltd., with money coming from S. C. Whitehouse, a Harrow garage proprietor, Nigel Holder and Straker of Messrs Kensington-Moir and Straker, who were Aston Martin distributors. Percy Kidner, who had been joint managing director of Vauxhall, prior to the General Motors take over in 1925, also provided support. In 1931 there was a brief flirtation with H. J. Aldington of Frazer Nash and later in 1932 with Lance Prideaux-Brune, another distributor. Stability came later that year when Sir Arthur Sutherland, a Newcastle shipping magnate bought the company outright, his son, R. Gordon Sutherland, sharing the joint managing directorship with Bertelli. During these years a range of good-looking sports cars was produced, powered by the faithful 1.5–litre overhead-camshaft unit, with competition successes being achieved at Brooklands, Le Mans and the Mille Miglia.

Unfortunately the Bertelli/Sutherland relationship proved a controversial one, the former leaving in 1936 and, though the 2–litre cars of the following year superficially resembled their forebears, they somehow lacked the earlier magic. Claude Hill, whose association with Bertelli went back to R and B days at Tyseley, took over as chief designer and was responsible for the Atom saloon, designed in 1939. The Atom — it was 'lots of power in a small package' said Sutherland — was powered by a 2–litre overhead-camshaft engine, the first of three power units. The final version was a 2–litre pushrod engine and, during the war, Hill and Sutherland covered 100,000 miles in this prototype.

Aircraft components were produced during hostilities and after the war the company, although in good shape, didn't possess the resources to develop the new design. Aston Martin was therefore put up for sale and bought early in 1947 by David Brown, who had seen an advertisement for a sports-car company in the pages of *The Times*. Brown was chairman and managing director of David Brown and Sons (Huddersfield) Ltd., gear manufacturers, who had diversified into the tractor market before the Second World War. Brown paid out £20,000 for Aston Martin which at the time consisted mainly of the Atom saloon! Then, the following year, he purchased Lagonda, after bids had been received from Armstrong Siddeley, Jaguar and Rootes. Lagonda, like Aston Martin, had produced a prototype design for the postwar market and from the assets of these two companies sprang the DB (for David Brown) Aston Martins.

Lagonda had been founded by an American, Wilbur Gunn, at the turn of the century, the name being borrowed from his brother-in-law's business back in Springfield, Ohio and was the Shawnee Indian name for the local Buck's Creek. The business started off in the back garden of Gunn's house in Staines, Middlesex and cars were produced from 1906. After chasing the mass-production market, the company consolidated a sporting tradition with the 2–litre of 1928 but finances were never sound and Lagonda went into liquidation in 1935. Fortunately, lawyer Alan Good purchased the company for £67,500 pipping Rolls–Royce to the post. W. O. Bentley was brought in as technical director and the range was rationalized. The outcome was the Meadows-engined LG6 and V12 models of 1938, Bentley and Stewart Tresilian being responsible for the latter design. The well-proportioned and handsome bodywork was styled by Lagonda's Frank Feeley.

It soon became clear that the postwar market was going to be rather more austere than the pre-war one, so a new, much smaller, car was planned during hostilities. The engine was a 2.6–litre, twin-overhead-camshaft six designed by Bentley and mounted in a truly cruciform chassis. Suspension was independent all round: coil and wishbone at the front with semi-trailing arms sprung by longitudinal torsion bars at the rear, though inboard brakes were an unusual fitment. Rack and pinion steering was employed. Prototypes were built, Feeley designing saloon and drophead coupé bodies. The car was announced, early in 1947, but Alan Good had rather lost interest in car manufacture by this time. Therefore, David Brown was approached by Tony Scratchard, the Lagonda distributor in Bedford, who urged him to buy the company. Brown eventually paid £52,500 for it, though this did not include the Staines factory so production was concentrated on the Aston Martin works at Feltham.

David Brown was certainly impressed with the Atom but felt that an open model was more in the Aston Martin tradition. The first car produced under his aegis, therefore, was a 2–door convertible, powered by Claude Hill's 1970cc pushrod engine, built in small numbers between September 1948 and May 1950. Retrospectively known as the DB1, the model had a multi-tube chassis, based on the Atom, but with coil-spring and trailing-arm suspension in place of the cantilever type used earlier. A racing version of this model won the 24-hour race at Spa in 1948, setting the pace for the marque's successes during the following decade. The following year, three special cars were developed for racing, two were fitted with Claude Hill's pushrod engine while the third was powered by the 2.6–litre twin-cam Lagonda unit. This union of Hill's chassis and Bentley's engine was, in effect, the prototype DB2 but Claude was naturally disappointed that a 6–cylinder version of his 2–litre wasn't going ahead and he left the company in 1949.

However, the DB2 went into production in closed and drophead coupé form, Frank Feeley being responsible for the styling. The model was available from May 1950 with a more powerful Vantage version being sold from January 1951. Production ceased in April 1953, but in October of that year the DB2/4 was announced with

What the other drivers saw. A DB4 pictured at the rear of Aston Martin's Newport Pagnell factory in 1958. This was the company's first postwar model where the boot did not connect with the passenger compartment

increased passenger accommodation. Under the bonnet was a more powerful 2.9–litre version of the twin-cam six (achieved by staggering the bores), this capacity being standardized from thereon. This model, with bodywork by Mulliners of Birmingham, was dropped in October 1955, and replaced with a Mark II version, with Tickford of Newport Pagnell responsible for the bodywork, David Brown having purchased that coachbuilding concern in 1955. Production continued until August 1957, though March of that year saw the DB Mark III with front disc brakes listed as a first-time option on an Aston Martin. Again, a high performance engine was available, in this instance from 1958, until the model was phased out in July 1959.

We should not forget that, at this time, Lagondas were also being produced (albeit in much smaller numbers) at Feltham. The 2.6–litre model went into production, early in 1949, offered with a range of Tickford coachwork, a Mark 2 version appearing with higher compression ratio at the end of 1952. The engine's capacity was increased to 2.9–litres in 1953, foreshadowing the DB2/4 Mark II, with bodywork re-styled and in 1957 a Mark 2 3–litre appeared, featuring a floor gear change and other minor modifications. The convertible version was dropped too, with 2– and 4–door saloons continuing until the model was discontinued in February 1958. The Lagonda name was briefly revived in 1962–1964, with the DB4-based Rapide 4–door saloon with automatic transmission and de Dion rear axle.

Aston Martin racing activities were consolidated with the appointment in 1950 of John Wyer as competitions manager. (In 1955 he became technical director, and general manager the following year, a position he held until 1963.) The arrival of the Eberan von Eberhorst-designed DB3 sports racer of 1951 showed that the company were in earnest and this was succeeded in 1953 by the DB3S. 1959 was the company's golden year when they became the only British firm to win the Sports Car Constructors Championship, the high spot being the Shelby/Salvadori DBR1 taking the chequered flag at Le Mans with another DBR1 car in second place. It was a good note on which to bow out, although a few 'Project' GTs were made in 1962 and 1963.

A new generation of sports cars was introduced with the announcement of the DB4 in October 1958. It was a superbly-proportioned, close-coupled coupé styled by Touring of Milan and manufactured using their Superleggera method of

body construction. The chassis, designed by Harold Beach, was of the platform type fitted with coil-and-wishbone independent front suspension while coil springs and radius arms, located by a Watts linkage was used at the rear. Servo-assisted Dunlop disc brakes were fitted all round and rack and pinion steering was employed. The engine was an entirely new unit for a production Aston Martin — an all-aluminium, twin-overhead camshaft six designed by Tadek Marek. It was of 3670cc (92 × 92mm) and developed 240bhp. The interior was beautifully trimmed in leather and the car cost £3976.

A GT-labelled version was also available with a 12-plug cylinder head and Weber carburettors.

Off for the weekend well loaded with luggage . . . A 1955 DB 2/4 sports saloon that would have cost its owner £2728

Other differences between this and the standard car were Girling disc brakes, Borrani wheels and a limited slip differential. Most examples were fitted with a 30-gallon fuel tank. Wheelbase was 3-inches shorter than standard. The Zagato bodied DB4 GT was built concurrently but with a higher compression ratio.

In 1960 there came Series 2 modifications to the basic model. These included larger front discs ($11\frac{1}{8}$–$12\frac{3}{8}$ inch diameter) and sump capacity increased from 14 to 17 pints. A new-style bumper and chrome surrounds to the windows were introduced. In April 1961, sump capacity was again increased, this time to 21 pints, and other changes included a better heater and improved windscreen wipers. In addition, a Laycock overdrive unit became an optional extra. In September details of the Series 4 DB4 were revealed. The main changes were a twin-plate clutch and the addition of an oil cooler. The radiator grille was modified while the bonnet scoop was lower and smaller. The following month a drophead coupé version was announced.

Another high-performance version of the DB4 engine appeared early in 1962. These Vantage, or Special Series, cars carry the SS suffix to their chassis numbers. The compression ratio was raised

Rare luxury. A DB5 drophead coupé of 1964 of which only 125 were made. A steel hard top was an optional extra

from 8.25 : 1 to 9 : 1 with output boosted to 260bhp. The DB4's headlamps were also faired-in at the same time, following the style of the GT. For the 1963 season the Vantage was offered with a GT engine, and the model was known as the Vantage GT. However, the front brakes of this Series 5 model were smaller, and the rear wings longer. The DB4 range was discontinued in June 1963 and replaced in October by the DB5.

This model had a larger capacity 3995cc (96 × 92mm) engine, which developed 282bhp and closed and convertible models were announced simultaneously. Electric windows were a refinement and Borg–Warner automatic transmission was offered as an optional extra (a few DB4s had been so equipped). For the first few months of production overdrive was fitted but from December 1963, a ZF 5–speed gearbox was substituted. The model will probably be forever identified with the special car produced for the James Bond film *Goldfinger*, of which one actual car and two replicas were made each costing about £15,000.

It was during 1964 that Aston Martin left Feltham, which had been their home since the 1920s. From thereon all production centred on the former Tickford Works site at Newport Pagnell, engine assembly and bodywork having transferred there some years previously.

Later that year, in October, a Vantage-engined version of the DB5 appeared with triple Weber carburettors developing circa 314bhp. Early in 1965 modifications to the DB5 included the fitting of a 10-inch diaphragm clutch and handles to the electric windows! A few estate cars, by Harold Radford, were also produced during the model's production span which ceased in October 1965.

It was replaced by the DB6, basically similar to the earlier model but with a long wheelbase (8ft 5¾in instead of 8ft 2in) which improved rear passenger accommodation. The spoiler lip at the rear is a distinctive feature. A Volante convertible had been announced in 1965 but this was on a DB5 chassis and when the longer DB6 version appeared the earlier model became known as the shortwheelbase Volante. A Mark 2 version of the DB6 appeared in July 1969 with larger wheels, flared wheel arches and re-shaped back seats, production ceasing in November 1970.

In the meantime the DBS had been announced in September 1967. As Touring had closed down their coachbuilding operations the new model was styled by Williams Towns and was designed to accommodate a new 5.3–litre V8 engine, then still under development. So the engine was much the same as the DB6 but a de Dion rear axle was used. However, the V8 eventually took over in April 1970. The last model to use the six was the Vantage version of the DBS, the final one leaving Newport Pagnell in July 1973.

Aston Martin production 1948–1959

DB1	15
DB2	409
DB3	10
DB2/4 Mark I	565
DB3S	31
DB2/4 Mark II	199
DB Mark III	551

DB4 1959–1963
DB5 1963–1965
DB6 1965–1969

(For specifications and development history see previous text.)

Check points

Body and chassis

All these cars employed the Superleggera method of construction which produced a strong, rigid, but

The longer wheelbase DB6 of 1966. The spoilered tail was a distinctive feature. Mechanically the same as the DB5, except for spring ratings, the DB6 cost £4998

light structure. The aluminium panelling doesn't suffer much from corrosion, with the exception of the doors which are vulnerable along their bottom edges, a condition usually caused by blocked drain holes. Check whether the electric windows are working — if they're fitted. They were specified for the DB5 and 6 and the motors are notoriously troublesome. Unfortunately, the platform chassis rusts badly at the bottom of the scuttle and to reach it requires body surgery. The other problem you'll probably experience is corroded jacking posts. Look for slight bulging of the metal inside the rear wheel arches and remember rust can also spread to a nearby side-member.

Suspension, steering and brakes

The independent front suspension has no particular vices, but the kingpins are rubber gaitered and, if these split, water can enter and accelerate wear on the ball joints and thrust pads. At the rear end check the condition of the back axle's parallel trailing link mountings as they can corrode. The rubber bushes must be changed every 60,000 miles.

You can gauge the general condition of the suspension by pressing down on the front wing. The car shouldn't move at all but do be careful how you press. Use the flat of your hand with the other one on top of it, otherwise you risk denting those elegant but delicate aluminium panels.

The steering rack is rubber-mounted on all cars except those fitted with power-assisted steering and should be checked for gaiter wear. As the mountings can perish, it is worth checking periodically that all is well. A tyre lever strategically placed between the rack and the chassis will soon indicate excessive movement.

Most cars employed Girling discs all round with twin circuitry and double-servo units. The exception was the DB4, which used Dunlop discs assisted by a Lockheed servo. The later brakes were a great improvement on the Dunlops, which had a reputation for rapid wear coupled with a need for a muscular right leg!

Engine, gearbox and transmission

Oil pressure is a vital factor when considering the state of the Aston Martin's twin-overhead-camshaft, all-aluminium, 6–cylinder engine. Up until the Series 3 DB4 of 1961 an acceptable figure would be 70psi (hot) at 3000rpm while the other cars should have a reading of at least 95psi (hot) at the same revs. Should you have any doubts then the oil pressure gauge reads 0–100psi on the earlier cars, while the later variants are graduated 0–160psi. It's because the sump capacity started off at 14 pints and progressively increased until 21 pints was standardized. This particular modificaton was prompted by overheating and the advantages of these later engines are obvious, though it should be stated that many of these earlier cars have been updated. Also, engines built up until 1962 suffered from piston-ring problems and excessive oil consumption resulted. However, it is unlikely that an unmodified example will be encountered. Oil changes are normally recommended at 2500 mile intervals. An Aston Martin cannot be run on a shoe-string!

A useful pointer to the engine's condition is the state of the 6-liner bleed holes in the crankcase on the carburettor side of the block. They shouldn't exude anything, but if they are gently weeping water this indicates that coolant has penetrated the usually empty gap between the upper neoprene sealing ring and its lower counterpart which are fitted to the lower end of the cylinder liners. The problem can be alleviated by adding Bar's Leak to the cooling system but it will only work if it hasn't been tried before. If this course of action doesn't have any effect, it's time for a specialist to look at the engine as it means that only the bottom seal is preventing water in the block from mixing with oil from the sump. Obviously the general condition of the engine is a crucial consideration if you're contemplating a used example as repair work isn't cheap. A power unit can be given a new lease of life for about £1500 at the time of writing while a total engine overhaul costs between £3000 and £4000.

Another important consideration is the state of the timing chains. There are two of these, one driving the camshafts from an intermediate gear and the second running from this gear down to the crankshaft. Yet a third chain drives the externally-mounted oil pump from the front of the crankshaft. It is customary for all three to be changed when a de-coke is carried out, and this is specified at 60,000-mile intervals. It's worth noting that valve clearances are achieved either by grinding the ends of the valve stems or changing to different-sized tappet cups.

As you may have gathered, working on one of these engines cannot be contemplated by an amateur, highlighted perhaps by the business of building up the bottom end. The shaft's seven main journals have to be fitted with their correct size bearings which are available in three grades and colour coded black, red and green. Crank grinding is possible to 0.010in and 0.020in. Then the chrome vanadium cylinder liners are shrunk into the block and the only way to remove them is to steam them out. . . .

The 10-inch diameter single-plate Borg and Beck clutch fitted to the DB4 cannot be regarded as

Interior of the DB6. The instruments were larger than the earlier cars. *Below*, the model's 325bhp Vantage engine with triple twin choke Weber carburettors

the model's most outstanding feature as it is heavy to operate and quick to wear. The DB5 featured a Laycock diaphragm unit, though the Borg and Beck version used on the DBS is reckoned to be the best and the earlier cars, with minor modifications to the release mechanism, can be so improved. Again the DB4 suffered from gearbox trouble caused by the synchromesh wearing badly. In extreme instances the cones can tip off their saddles preventing gear engagement. This particularly alarming shortcoming manifests itself only when the box is really hot and is unlikely to show up on a quick trip round the block. The Series 3 DB4 of 1961 saw an overdrive option introduced but from the DB5 onwards a five-speed ZF unit was fitted and, though it had a reputation for being unacceptably noisy when new, its internals have stood the test of time.

Borg–Warner automatic transmission was introduced as an option on the DB5 and this might be a variant to avoid as the engine's low-speed torque didn't take too kindly to such fettering. It depends what you're looking for. There were certainly more DB6s so equipped and you're more likely to encounter this model than DB4s or 5s because it was in production at the time of Aston Martin's sensational £1000 price cut.

Rear axle problems are rarely experienced.

Spares

Don't forget that you can still obtain parts of these 6–cylinder cars from the factory though this applies, in the main, to mechanical components. Other spares and services can be obtained from the dealers listed below. It is well worth joining the Aston Martin Owners' Club as it has an excellent magazine and caters for all Astons.

DB4/5/6 spares and services

Aston Martin Lagonda (1975) Ltd., Tickford Road, Newport Pagnell, Bucks (tel: 0908 610620).
H. W. Motors Ltd., New Zealand Avenue, Walton-on-Thames, Surrey (tel: Walton-on-Thames 20404).
Robin Hamilton, Fauld, Tutbury, Burton-on-Trent, Staffordshire (tel: Burton-on-Trent 813939).
Hyde Vale Garage, Hyde Vale, Greenwich, London SE10 (tel: 01-692 2822).
Richard Stewart Williams Ltd., 31/35 Padfield Road, London SE5 9AA (tel: 01-733 1062/0659).
Aston Martin Owners' Club Ltd. Gerry Hopkins, 293, Osborne Road, Hornchurch, Essex (tel: Hornchurch 46841).

Performance and consumption (DB5), *Motor*, February 6, 1965.
Top speed: 145.2mph. Overall mpg: 17.6 for 1960 miles.

Aston Martin DB4/5/6 production

DB4 Series 1		150
DB4 Series 2		350
DB4 Series 3		164
DB4 Series 4		270
DB4 Series 5		185
	TOTAL	1119
DB4 GT		75
DB4 GT Zagato		25
DB5		1150
Volante		37
DB6 (all)		1330
DB6 Mark 2 (all)		245
Volante		140
Volante Mark 2		38
DBS*		787
AM Vantage*		70

*Also fitted with 4–litre twin-cam 6–cylinder engine.

Austin 1906 to present day

Herbert Austin was 38 years old when he lost his job as general manager of the Wolseley Tool and Motor Car Company and set up on his own at Longbridge, Birmingham in 1905. Production began the following year and 4–cylinder models were initially offered although sixes were available from 1908 to 1913. In 1909 a 7HP car was introduced, this single-cylinder model also being offered as a Swift. Longbridge's best-seller at the time was the 22.4HP Twenty, a smaller version, the Ten, appearing in 1910. By the outbreak of the First World War in 1914 there were three basic chassis available: Ten, Twenty and Thirty horsepower cars with bodywork also produced on the premises, an unusual state of affairs for the day.

After the war Austin dropped all his other models and perpetuated the best-selling theme of the Twenty with a re-designed 3.6–litre 4–cylinder car in 1919. Unfortunately, the new car didn't catch on and the unsettled financial climate forced the hasty production of the 1.6–litre Twelve in 1921. This hard-wearing and phenomenally-reliable car remained available until 1936, though it was increased in size and engine capacity (1861cc) in 1927. In taxi form it remained a charming ingredient of the London scene until the 1950s.

In 1922 came the Austin Seven, probably the most famous British car of the inter-war years. The story of its creation has already been recounted but this 'baby' and the Twelve formed the basis of Austin's survival in the '20s.

A fashionable 6–cylinder car, the 3.4–litre Twenty appeared in 1927, though for a time the older four remained available. In 1928 came a smaller version of the 6–cylinder Twenty — the 2.3–litre Sixteen which initially shared the same body and chassis as the Twelve. A less happy confection was the Twelve Six of 1931. However, the Ten of 1932 was well suited to the times and, in updated form, remained in production until 1947. In 1933 came the 1535cc Light Twelve, a theme that again lasted into the postwar years. The Seven's replacement, the 900cc Eight, went into production in 1939 and that year saw the range of models receive a much-needed facelift with updated styling and alligator bonnets. Other models available at the outbreak of the Second World War was a range of big sixes of Fourteen, Eighteen and Twenty-Eight horsepower.

During the Second World War, the Longbridge works and shadow factory close by at Cofton Hackett produced bomber and fighter aircraft together with 120,000 vehicles including trucks, ambulances and even Austin Eights!

Not surprisingly, the company was quick off the mark with its postwar programme and announced the first 'new' model in September 1944. The Austin Sixteen was, in fact, the Austin Twelve intended for the 1940 season fitted with a 2.2–litre overhead-valve, 4–cylinder engine (a first for a Longbridge car) developed during the war. The rest of the range was rather dated though, in effect the Eight (Series AS1), Ten (GS1) and Twelve (HS1) models that had been announced in 1939.

The first really new design of the postwar years was the A125 semi-razor-edge Sheerline (DS1) of 1947 powered by a 4–litre engine. The independent front suspension was another first for an Austin car. The taxing of cars by RAC horsepower rating was phased out at the end of 1946 and the new cars from Longbridge reflected in their titles, the approximate brake horsepower developed by their engines. Therefore, the new overhead-valve 1200cc saloon, announced in October 1947, to replace the Eight, Ten and Twelve, was designated the A40 and was available in 4–door Devon (GS2) and 2–door Dorset (G2S2) form.

Then also in 1947 was the A135 Princess (DS2) and later in longwheelbase form (DM2). These had similar mechanics to the earlier Sheerline but with triple SU carburettors instead of a single Stromberg unit. Aluminium coachwork with faired-in headlamps was by Vanden Plas, Austin having absorbed the famous coachbuilders in 1946. Both models were initially announced in 3.5–litre form but this soon gave way to the bigger capacity.

From thereon models came thick and fast. In July 1948, the A70 Hampshire six light saloon (BS2), powered by the Sixteen's engine was announced, but that model was phased out the following year. Early in 1949 came the larger-bored, 2.6–litre, A90 Atlantic drophead coupé, with power-operated hood for £30 extra and clearly geared for the American market. A year later a short-lived Atlantic saloon (BE2) was announced. The 1951 Hereford A70 four light saloon (BS3) and drophead coupé (BD3) replaced the Hampshire, while a sports version of the A40 with bodywork by Jensen Motors (GD2), appeared at the same time. A column change A40 Sports (GD3) was produced in 1951–1952.

A new small car, the spiritual successor to the Austin Seven (early examples were so named), made its debut at the 1951 Motor Show. This was the A30 4–door saloon (AS3), which was Longbridge's first foray into integral body–chassis construction. The engine was a new, overhead-valve 800cc four, designed on similar lines to the A40 1200cc unit of 1947. The A40 Somerset, a replacement for the A40 Devon and Dorset, appeared in 1952, and was available in 4–door saloon (GS4) and coupé (GD5) form.

New models for 1955 included yet another A40 face lift — the Cambridge — available with 1200cc engine (GS5) or in A50 form with 1489cc (HS5) power unit. Simultaneously, the similarly-styled A90 Westminster (BS4), to replace the A70 Hereford, was announced, as a 2.6–litre six.

A high-performance version of the A90 appeared in summer 1956. This A105 was fitted with a twin-carburettored version of the new six (BS5), though there were minor body changes and a change in compression ratio (BS7) within 6 months. At the same 1956 Motor Show the A90 became the A95 (BS6). Although the Sheerline had been phased out in 1953 the Princess soldiered on, was re-styled for 1957 and automatic transmission standardized. The A30 was fitted with a 948cc engine for 1957 to become the A35 (AS5).

Early in 1957 a higher compression version of the A50 was announced, thus creating the A55 (HS6). During the year the Nash Metropolitan, which had been produced by Austin for American Motors for the US market, became available on the British market as a plain Metropolitan. It remained with the 1489cc A50 engine until 1961.

Austin's answer to the Land–Rover, the Gypsy, went on sale in 1958 using the faithful 2.2–litre Sixteen engine and with all-independent Flexitor rubber suspension. But the really significant event of the year was the arrival of the first 'new' car to appear since the creation of BMC in 1952. The 948cc A40 featured a completely new 2–door body (AS6) designed by Farina with a Countryman version (AW6) appearing the following year.

A new big saloon, also styled by Farina, was announced early in 1959. This was the A55 Cambridge (HS8) and later the same year the A99 Westminster (BS9), powered by the 3–litre 6–cylinder engine, was produced. The great event of 1959 was the arrival of Alec Issigonis's revolutionary front-wheel drive Mini, Longbridge versions being named Austin Seven, until 1961, in deference to its famous forebear. From thereon the marque became almost solely front-wheel-drive.

An upmarket version of the big Farina saloon, the 3–litre Princess, went on sale in 1960. Produced by Vanden Plas, and sold under this marque name, from 1964 a Rolls–Royce-engined version replaced it. Named the Princess 4–litre R, a 3909cc aluminium, overhead inlet/side exhaust power unit was fitted with automatic transmission and power steering as standard.

A twin-carburettored version of the Mini, the 997cc Cooper version arrived in 1961 with the hotter S variant in 1963. The next development of the front-wheel-drive concept came with the arrival, for 1964, of the Austin version of the 1100 saloon that had borne the Morris badge since its introduction in 1962. From 1965 the Hydrolastic suspension was fitted to the Mini range. Yet another, similarly suspended, front-wheel-drive offering was the sturdy 1800 of 1965.

The A40 Sports model, produced by Jensen for Austin between 1950 and 1952

On a more conventional front the Farina saloons continued in production for most of the decade. A Mark II version of the A40 with 1098cc engine was introduced for 1962, and was available until 1967. Also in 1962 came the 1622cc-engined A60 which was on sale until 1969. The A99 became the A110 Westminster in 1962 utilizing the same body and remaining in production until 1968. The same year the 4-litre Princess was phased out.

Short run Austins 1947–1968

A40 Sports 1950–1953	3800
Princess I 1947–1950	800
Princess II 1950–1953	760
Princess III 1953–1956	350
Princess IV 1956–1959	200
Princess A135 long wheelbase 1952–1968	3344
A105 1958–1959	500
3 litre Princess Mark I 1959–1961	4719
3 litre Princess Mark II 1961–1964	7984
Princess 4-litre R 1964–1968	6555
Sheerline 1947–1950	
DS1 short wheelbase saloon	8000*
DA1 limousine long wheelbase DA1 ambulance DH1 hearse	1000*
A90 Atlantic 1948–1952	10,000*

*approximate figure

A30 1952–1956
A35 1956–1962

Specifications and developmental history

The A30 Seven saloon was announced at the 1951 Motor Show. It was a 4-door monocoque with coil-and-wishbone independent front suspension and a Cam Gears cam-and-lever steering box, though the Burman worm-and-nut unit was an alternative fitment. Lockheed hydraulic brakes were employed, though the rear ones were operated mechanically via a hydraulic cylinder — a curious arrangement. The model marked the appearance of what was to be designated the BMC A series engine. This 800cc (58 × 96mm) 4-cylinder power unit had pushrod-operated overhead valves and developed 30bhp. The interior seats were upholstered in PVC material, the rear ones being individually moulded. The price was £554.

A 2-door version was introduced at the 1953 Motor Show having a re-designed facia with centrally-grouped instruments. This also applied to the 4-door (AS4). In November 1953 the seating was revised, with increased head and knee room, while the spare wheel was mounted vertically in the boot to allow greater capacity. In April 1954 the rear axle ratio was changed from 5.125:1 to 4.875:1. September 1954 saw the appearance of the Countryman version (AP4). This and the saloon versions were phased out in September 1956.

The model was re-designated the A35 for 1957 being fitted with a 948cc (63 × 76mm) version of the A series engine developing 34bhp. It was available in 4-door (AS5) and 2-door (A2S5) form. The main external differences were a painted radiator grille with chrome surround replacing the plated grille, larger wrap around rear window and flashing light indicators in place of the semaphore units. Internally, a remote control gear change was fitted. A Countryman version (AP5) appeared at the same time. The saloon models ceased production in July 1959 but the Countryman remained available until September 1962, was slightly restyled in March of that year (AP6) and flashing indicators introduced.

Check points

Bodywork

The most vulnerable areas on these models are the sills and inner wing panels. Both the inner and outer sills tend to suffer, the damage caused by water and mud thrown back by the front wheels. The nearside usually suffers to a greater degree — the legacy of many a kerbside puddle. If the rust really attacks the metal, the corrosion can spread to the outer edges of the floor pan, the front inner wing and also to the forward section of the inner rear one. Less significant structurally, though

nonetheless worthy of examination, are the front wings. Check particularly around the headlight recesses and the body seam below the light unit. Be prepared for wear in the door hinges while front and rear screen rubbers may be admitting water.

Suspension, steering and brakes

The independent front suspension layout incorporates a double-acting Armstrong lever-type shock absorber with the top wishbone. Wear in this hard-working unit results in a certain amount of front-end wallowing but reconditioned replacements are available. Another shortcoming is caused by the king pin layout. The pin is cotter-pinned on to a threaded pin which passes through the wishbone eye. Unfortunately, this threaded pin does have a habit of turning and burrowing into the wishbone itself. The only practical solution is to get another wishbone, though re-bushing is a possibility. At the back of the car the reverse-camber semi-elliptics aren't renowned for breakages, though the shackles would repay inspection.

Steering is largely trouble free, though the box's bottom seal does leak to the detriment of its lubricant thus accelerating wear.

The brakes were not the model's strongest point, relying on 9-inch drums. A surprising amount of pressure on the pedal may be needed to arrest the car's progress!

Below, left, **the four door Austin A30 in its original 1952 guise.** *Below*, **larger wrap around rear window and flashing indicators in place of semaphores, indicates this 1956 A35**

The A30's engine in its original 803cc form in 1952. Later designated the BMC A series unit, its derivatives are still in production at the time of writing (1980)

Engine, gearbox and transmission

The BMC A series engine is a thoroughly reliable unit evidenced by the fact that it is, in essence, still with us to the present day. This engine will continue running without too much protest even when it is in an advanced state of wear. A marked decline in performance may be the result of a burnt out exhaust valve, which occasionally occurs. In general, there is little to worry about as parts are universally available. A minor shortcoming of the A30 engine is that it was fitted with a bypass oil filter while the A35 boasted a full flow unit.

Similarly, the gearbox on the A35 is better than

the one fitted to the earlier model, that having Austin Eight origins! The rear axle has a reputation for long life and reliability.

Spares

The engine parts are also common with the Morris Minor and Austin Healey Sprite and the same applies to the independent front suspension system in the case of the latter model. New bodyparts are unavailable unless you stumble on an Aladdin's Garage. Otherwise try the Owners' Club.

Austin A30–35 Owners' Club, Bill Cochran, 25 Fiveheads Road, Horndean, Portsmouth, Hants.

Performance and consumption (A35 2–door), *The Motor*, November 21, 1956.
Top speed: 71.9mph. Overall mpg: 37.7 for 999 miles.

Basic but functional, the driving compartment of a 1953 A30. Balanced drop windows and therefore no winders, were fitted

A30 and A35 production by chassis number

MODEL	IDENTIFICATION CODE	CHASSIS NUMBER FROM	TO
A30 Seven Saloon			
May 1952–November 1953	AS3	101	30602
Seven Saloon			
October 1953–September 1956	AS4	30715 LHD 25621 RHD	224327
	A2S4	25631 RHD 33906 LHD	224326
Countryman			
September 1954–September 1956	AP4	73987 RHS 89121 LHS	223678
A35 Saloon			
September 1956–July 1959	AS5	120 LHS 112 RHS	203676
Saloon			
September 1956–July 1959	A2S5	103 LHS 105 RHS	187589
Countryman			
October 1956–January 1962	AP5	3291 RHS 827 LHS	277159
Countryman			
March 1962–September 1962	AP6	2101	10954

Austin Healey 1953–1971

Donald Mitchell Healey (born 1898) was a single-minded Cornishman from Perranporth. His father, Fred, ran the local general store and was also an active building contractor. After education at Newquay College, Donald moved east to the Sopwith Aviation Company at Kingston-on-Thames, Surrey, his father paying for an apprenticeship, as there were few prospects in Cornwall at the time. However, the First World War broke out and before he completed his time with Sopwiths, the young Healey volunteered for the Royal Flying Corps but was invalided out in 1917, after his plane had been hit by allied anti-aircraft fire in France. From thereon he worked for the Aircraft Inspection Department.

After the war, Donald returned to Cornwall and in 1919 he built the Red House Garage, next to his father's shop in Perranporth. Initially, he became a keen competitor in local competitive motoring events, and in 1924 he entered the London MCC trial, but his entry of a Riley Redwing was damaged by fire *en route* to London. He immediately returned to Cornwall, collected his trusty ABC, competed in the event and returned home with a gold medal. In 1929 he looked farther afield, entering the Monte Carlo Rally but failing to qualify. The following year he was back again, finishing seventh in a supercharged Triumph Super Seven. For the next three years he drove Invictas in the Monte, hitting the rallying jackpot with a win in 1931 and a second place in 1932.

In 1931 he joined Riley at Coventry moving on, in 1933 to Triumph, initially as experimental engineer, soon after being promoted to technical director. Monte Carlo and straight eight Dolomite were just two models produced under his aegis at Triumph. Involvement in rallying increased. In 1934 Healey drove a Triumph to third place in the Monte Carlo Rally, while the marque achieved a second placing in 1935, though he crashed in Denmark in a straight eight Dolomite, fortunately without injury to himself or his passenger. He moved to Lucas, briefly, in 1937 but returned to Triumph as a member of the board later the same year. Unfortunately, the firm went into liquidation in 1939 and was purchased by Thomas Ward, a Sheffield-based engineering and steel-making group. Healey stayed on as general manager, negotiating the sale of the old Gloria works to H. M. Hobson, the carburettor manufacturer, where he acted on behalf of the Air Ministry. After a short stay with the new incumbents, Healey moved to Humber and there became concerned with armoured car development.

It was during his spell at 'the Humber' that Healey's ideas about producing a car crystallized in 1943–1944. He met with three individuals who shared his enthusiasm for the project, and the intention, at this early stage, was to sell the idea to the moribund Triumph company in the shape of Thomas Ward Ltd. The unofficial design team was made up of Achille 'Sammy' Sampietro, an Italian, who had worked for Alfa Romeo and Maserati before coming to England, Ben Bowden, an ex-Farina stylist who was then a body man with Humber, while James Watt was a salesman whom Healey had met at Triumph and was also a fellow member of the Royal Air Force Voluntary Regiment.

Watt had a couple of fairly successful meetings with Thomas Ward but at the final one in February 1944 he was told that Wards had decided not to back the project as the team were not car manufacturers. Up until then the un-named vehicle had been temporarily called 'the Triumph' but after the Ward rebuff it simply reverted to 'the car'. At this point it was decided to push ahead with the design, relying on the modest resources available. Part of this help was provided by yet another Rootes employee, Peter Skelton, director of the Humber distributor, Westlands, in Hereford, who offered to produce the chassis for the new car there. It was at Westlands that Sampietro noticed an 8-foot folding machine, a device that creased metal up to 8 feet long and the Healey chassis was, therefore, adapted to fit in with this dimension! Healey's old friend Victor Leverett, who worked for Riley, suggested that they use the company's 2.4–litre power unit. A meeting with Victor Riley was arranged and he offered one month's supply of engines in advance. Now all that was needed was somewhere to build the cars! Fortunately, a former Triumph director, Wally Allen, chairman of Benford Ltd., who produced concrete mixers from a factory at Warwick, offered a corner of his works for chassis assembly. The prototype Healey was completed there during 1945 and revealed to the Press early the following year. The chassis was fitted with trailing arm independent front suspension — Sampietro had been influenced by the

Volkswagen layout — but with coil springs fitted front and rear. The traditional Riley torque-tube transmission was also employed. This first car was fitted with open two-seater bodywork by Westlands of Hereford and a saloon by Elliot of Reading was also available. Both options were available until late 1950. Ben Bowden had sketched out these shapes on his dining room wall and then traced the originals. . . . When introduced, the Healey was the fastest production saloon in the world having covered the flying start mile at 104.7mph but it was expensive at £1598. As production built up, activities soon outgrew the Benford premises and a new works at The Cape, Warwick was established.

Minor chassis modifications were made as output progressed. The original chassis had been designated the A type, while the B of mid-1947 was fitted with an adjustable steering column, boot-mounted petrol pumps and single 12-volt battery. The slab-sided Sportsmobile model of 1949–1950, for export only, used a lengthened B-type chassis but achieved little success. The C type of 1950 had Elektron front suspension boxes with side plates, while the D- and E-type chassis were allocated to the Silverstone of 1949. This was a cost-conscious 2–seater selling for £1246 with bodywork by Abbey Panels and design by Healey's Len Hodges. The E type of 1950 had a slightly wider body, bench-like front seat and was easily identified by the bonnet airscoop. The Tickford saloon and Abbott drophead coupé, available from mid-1951, had triangular air intakes adjacent to the radiator grille and were fitted to the BT-type chassis. The F-type, which appeared in November 1951, used an open propeller shaft and hypoid rear axle instead of the Riley unit and torque tube, a modification shared with the RM saloons.

A chance meeting between Donald Healey and George Romney, general manager of the Nash–Kelvinator Corporation, on the *Queen Elizabeth* in 1949 had fortuitous repercussions for the Healey company. Romney was anxious to capitalize on the sports-car boom gaining momentum in America and the outcome was the Nash–Healey of 1950. This used a strengthened Silverstone chassis, designated N type, and was powered by a 3.8–litre, overhead-valve, Nash 6–cylinder engine. The car was styled at Warwick by Len Hodges and manufactured by Panelcraft, though from 1952 the body was re-styled and lowered by Farina and a 4.1–litre engine fitted. The model was for export only, but in 1951, the G-type Sports Convertible was announced, with a 3-litre Alvis engine replacing the Nash unit.

All models considered so far were made by the score and a few by the hundred, but the company's big break came in 1952 when their sensational 2–seater 100mph sports car was announced. Although the 1950 Nash–Healey sports car can't have earned much money in America, it did allow Healeys to build up their cash reserves and gain valuable experience of designing all-steel bodywork. At the end of 1951 Donald Healey and his son Geoffrey began work on the specifications of a new low-cost sports car with their sights set firmly on the growing American sports-car market. The intention was to fill the gap between the Jaguar XK120 and MG TD. Austin were approached and they readily supplied the 2.6–litre 4–cylinder engine, suspension and transmission of the then current Atlantic saloon. Healey's team then de-

Donald Healey's Westland Roadster, in which he finished ninth overall in the 1948 Mille Miglia. Still carrying its competition numbers, it was then road tested by *The Autocar*

veloped the model with Gerry Coker, who had joined the Healey company from Rootes in 1950, designing a handsome 2–seater body which, in some ways, was an extension of the earlier Sports Convertible format.

The car made a dramatic debut at the 1952 Motor Show, the blue left-hand-drive Healey 100 being offered at £850. BMC's Len Lord snapped up the design and 2 days later the car sported the *Austin* Healey badge. A new marque had been born, though the model's creation wasn't quite that spontaneous, as recounted in Part One! Lord reduced the price by £100 and the Austin Healey 100 went into production in 1953. Tickford had produced the prototype body but, as they weren't geared for quantity production, Jensen Motors undertook the construction of the bodywork with panels supplied by Dowty Boulton Paul. These shells were then supplied to Longbridge where the cars were finished. Completion was transferred to the MG factory at Abingdon in 1957, the model and its 6–cylinder derivatives being made until the end of 1967. Its demise was in deference to the American safety regulations that came into force in January 1968. 71,568 big Healeys were made, and the evolution of the model is traced in the next section.

The other string to Donald Healey's bow was the

The Austin Healey Sprite in its original 1958 guise with distinctive 'Frogeye' headlamps.

Sprite model. The car was conceived late in 1956 following a discussion between Healey and Sir Leonard Lord (as he became in 1954). Lord's idea was for a latter-day version of the Austin Seven Nippy and Ulster models. As it happened Austin had toyed with a 7HP 2–seater of their own in 1953, but it had not progressed beyond the paper stage. The new Healey was largely designed by Donald's son, Geoffrey, with Gerry Coker again being responsible for the styling. Production of prototype bodies (the first of which featured pop-up headlights) was assigned to Panelcraft, while mechanics were mostly borrowed from the A35 saloon. Known at Warwick as 'Tiddler', it was

Right, **the 1953 Austin Healey 100.** ***The Autocar*** **achieved a top speed of 111mph with this car**

given the Healey type number Q. The model was approved in 1957 and, though it was originally intended for Longbridge it was later decided to assemble it at Abingdon alongside its big brother. The car was launched in May 1958 and received instant acclaim, a distinctive feature being the 'Frogeye' headlamps.

The Sprite appeared in badge-engineered form as the MG Midget in 1961, the Austin Healey being phased out of production in 1971 after 129,354 had been built though the MG soldiered on until the end of 1979.

One of the last projects engineered between BMC and the Healey company was an intended replacement for the ageing 3000 model. This was the 3909cc aluminium, 6–cylinder, overhead inlet/side exhaust, Rolls–Royce engine, fitted at the time to the Vanden Plas Princess 4–litre R, and installed in a widened 3000 so that it could accept the Princess rear axle. The engine was lighter than the standard power unit and the performance impressive, even with the retention of the saloon's automatic transmission, but the project was dropped after the production of a few experimental cars. This was probably because the concept clashed with the E type Jaguar, the concern having come under BMC's wing to form BMH in 1966. This sealed the Big Healey's fate once and for all.

Healey production 1946–1952

A type	97
B type	231
Sportsmobile (B type)	(23)
C type	170
Silverstone	105
BT type	50
F type	100
Nash–Healey	253
Nash–Healey (Farina bodied)	151
Healey–Alvis (G type)	25
G type (others)	3

100 1953–1956
100 Six 1956–1959
3000 1959–1967

The Healey 100 was announced at the 1952 Motor Show. It was a 2–door open 2–seater, the chassis being a box-section structure with coil-and-wishbone independent front suspension and Burman steering box. Girling hydraulic drum brakes were employed. The engine was a 2662cc (87 × 111mm), 4–cylinder, overhead-valve push-rod unit developing 90bhp. The seats were upholstered in leather and the car priced at £1323. The Austin Healey 100 (Series BN1) went into production in mid-1953, minor changes being 11-inch brake drums (instead of 10-inch), modified steering gear and slightly raised headlights. The model was priced at £1063. Chassis numbers began at 138031.

From mid-1955 (from chassis number 228047) a Morris C series rear axle was fitted and later a 4–speed gearbox (BN2). In 1955 the Warwick-built 100S was available with aluminium body, smaller radiator grille, Weslake four-port cylinder head and Dunlop disc brakes all round. The 100M of 1955–1956 had an engine prepared to Le Mans specifications and louvred bonnet.

In 1956 came the 100/Six (BN4) powered by the BMC C series 2639cc (77 × 89mm), 6–cylinder overhead-valve pushrod engine developing 102bhp. Chassis numbers began at 62191. It is identifiable by horizontal slats in the radiator grille and bonnet scoop. It was a 2–seater with possible rear accommodation. Completion transferred to the MG factory at Abingdon in 1957 (from chassis number 50759). In 1958 a six-port cylinder head was introduced (from chassis number 101), the model designated BN6, bhp rising to 117. The Austin Healey 3000 appeared in 1959 with engine capacity increased to 2912cc (83 × 89mm), developing 124bhp. Chassis numbers began at 101, and 2-seater (BN7) and 4-seater (BT7) versions were available. Girling disc brakes were fitted on front wheels.

A Mark II version appeared in 1961 (from chassis number 13751) with triple-carburettored engine, developing 132bhp. The radiator grille was fitted with vertical slats. In 1962, the Mark II reverted to a twin-carburettored layout (BJ7), beginning at chassis number 17551. Other improvements were wind-up windows, a wrap-around windscreen and integral hood. This was replaced in 1964 by the Mark III version of the 3000 (from chassis number 25316), designated BJ8. Ambla upholstery was introduced with a leather option together with a wood veneer dashboard. Production ceased in 1967.

Check points

Body and chassis

Although the Big Healey has the appearance of being a rugged piece of machinery it is certainly susceptible to the rigours of rusting, and repairs can be expensive. The bodywork is mostly of steel, though the bonnet surround and its opposite number, the shroud around the boot aperture, are made of aluminium. Beginning at the front of the car, the state of the forward shroud should be examined as replacements can be expensive. Take particular note of the area beneath the radiator grille as this is a favourite rust point. The main problems at the front are confined to electrolytic action between the aluminium shroud and the steel panels attached to it. Look out for signs of bubbling paint immediately adjacent to the chromium beading. The junction between the front wings and the bulkhead can be troublesome and while examining this area check where the wing curves underneath the car as it may have come away from its mountings. Before leaving the wings, check also the area immediately below the headlamps — another rust spot.

The inner and outer sills should be the next area for scrutiny. Check the point of the inner one where it joins the floor and the outer ones which run between the door shut faces. Examine also the chassis outriggers for signs of corrosion. Although the outer sills and door shut faces are covered with aluminium finishing plates, this doesn't mean that the metal beneath is perfect. The door hinge pillars should be another area for investigation, especially below the bottom hinge.

Don't forget to look inside the car to examine the state of the floor. This tends to give trouble; suspect places are at the front, also adjacent to the inner sills and the points at which the rear wheel arches are evident. Another check point is the pan

Left, **the 1961 MG version of the Sprite theme, the 46bhp Midget, which was £40 more.** *Below*, **1956 100-Six with limited space for rear occupants**

Under the bonnet of a 3000 Mk II with triple SU 132bhp engine. A top radiator hose was normally fitted!

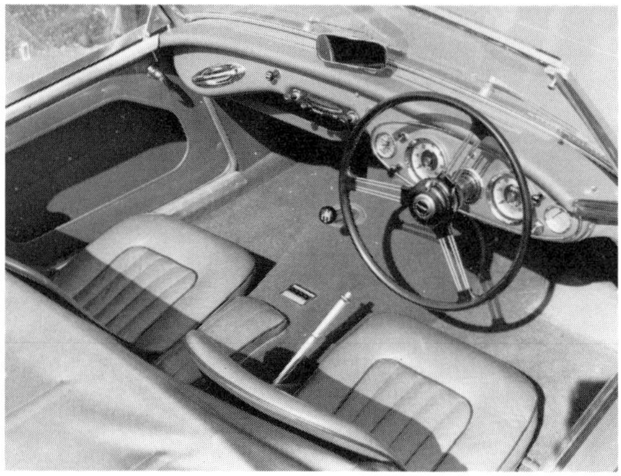

The 100's cockpit with metal dash

above the rear axle, or rather beneath the rear seats on those models fitted with them. Before continuing inspect the doors; they tend to rust underneath, while missing or cracked window glasses are difficult on the Mark IIA onwards as their double curvature represents a supply problem. Then on to the rear wings. Examine the area where they attach to the door sheet face. The familiar electrolytic corrosion occurs on the rear shroud so check the places immediately adjacent to the chrome beading. Also examine the inner wing. The boot floor is an important part of the structure, though it is difficult to inspect from the top because the petrol tank is in the way. The suspect areas are the seams and where the bumper dumb-iron support brackets run from the rear of the shroud to the floor. If these have rotted at their base they can prove difficult to replace. And if they're not accurately re-positioned, then the bumper will end up crooked. There may well be indications that the boot floor has given way in the past and if this is the case then the chassis outriggers that are attached to it will also be beyond redemption. The steel boot lid can also rust along its bottom edge.

The chassis should be checked around the front suspension pillar and adjacent to the front engine mountings. The main frame is fairly hardwearing but as we have seen, the same can't be said for the outriggers. Radius arms are employed to locate the rear axle on the Mark III cars and these are mounted on boxes alongside the rear wheel arches. Bump stops are fitted above the rear axle and these should similarly be checked.

Suspension, steering and brakes

The front suspension shouldn't present many problems though king pin wear is not uncommon. Rear springs are cheap to replace but they do have a reasonably hard life and tend to tire rather than fracture. You can assess their condition by measuring the distance between the tyre and the edge of the rear wing. The figure should be between 2 and 3 inches.

Steering gear is long lived and isn't a weak point. The same goes for the brakes, the front discs, fitted from 1959, being particularly reliable.

Engine, gearbox and transmission

The Healey's engines, both in 4– and 6–cylinder form are durable and reliable units. A smoking exhaust or evidence of oil forced out of the engine breather by crankcase pressure are signs of a worn engine and should be carefully examined. An acceptable oil pressure reading for the four is 45psi while the six should provide 50psi at over 2500rpm.

Gearbox problems are largely confined to the 6–cylinder cars which suffer from worn first gears and layshafts. Unfortunately, new spares ceased some time ago but AH Spares of Leamington Spa, Warwickshire do offer reconditioned layshafts and new first gears. The reconditioning is achieved by cutting off the first gear, boring out the remainder and pressing in a new cog and welding it back together.

Spares

Big Healey enthusiasts are fortunate that body and mechanical spares are available for these cars. A good source of parts is A. H. Spares and Southern Carburetters, addresses below. Restorations are undertaken by John Chatham Cars. The Austin Healey Club is an extremely active one and close contacts are maintained with the Healey family.

Big Healey spares and services

A. H. Spares Ltd., 427 Tachbrook Road, Whitnash, Leamington Spa, Warwickshire CV31 3DQ (tel: Leamington Spa 20477).
John Chatham Cars, 138, Gloucester Road, Bishopsgate, Bristol, Avon BS7 8NT (tel: 0272 44154).
Southern Carburetters, 49 The Broadway, Wimbledon, London SW19 (01-540-8128/2723).

Austin Healey Club
Eastern Centre, David Hicks, 102 Fairfax Drive, Westcliffe-on-Sea, Essex (tel: Southend 41670); *Midland Centre and Overseas*, Mrs Carolyn Walters, 27 Three Oaks Road, Wythall, West Midlands B47 6HG; *New Forest Centre*, Pat Martin, 104 Winchester Road, Shirley, Southampton (tel: Southampton 783680); *Northern Centre*, Sheila Reich, 61 Winstanley Road, Sale, Cheshire (tel: 061-973 9995); *Southern Counties Centre*, Phil Wakefield, 99 Bourneside Road, Addlestone, Surrey (tel: Weybridge 48698); *South Western Centre*, Carol Marks, 171 Coldharbour Road, Bristol; *Thames Valley Centre*, Tom Oakman, 14 Burnt Oak, Wokingham, Berkshire.

Austin Healey 3000 performance and consumption (Mark III), *Motor* March 28, 1964.
Top speed: 122.5mph. Overall mpg: 17.7 for 1301 miles.

Austin Healey 100, 100 Six and 3000 production

100	BN1	10,688
	BN2	3924
100S	AHS	55
100 Six	BN4	10,826
	BN6	4150
3000 Mk I	BT7	10,825
	BN7	2825
Mk II	BT7	5095
	BN7	355
	BJ7	6113
Mk II	BJ8	390
Phase two	BJ8	16,322
TOTAL		71,568

Sprite 1958–1971

Specifications and developmental history

The Austin Healey Sprite (Series AN5) was announced in May 1958. It was a 2–door, 2–seater convertible with a cheeky and distinctive 'Frog-eye' bonnet. Coil-and-wishbone independent front suspension was fitted with quarter-elliptic springs, while trailing arms were used at the rear. Lockheed hydraulic drum brakes were fitted and rack-and-pinion steering employed. The 948cc (62 × 76mm), 4–cylinder pushrod engine developed 48bhp. Seats were PVC-covered and the price was £678. Chassis numbers began at 501.

In mid-1961 a Mark II version (AN6) was announced, chassis numbers beginning at 101. This shared the same specifications as the MG Midget that went into production at the same time. The original front was replaced with a more traditional one and the engine's compression ratio was raised to 9 to 1 and bhp increased to 48.9.

For 1963 the Sprite was fitted with a larger 1098cc (64 × 83mm) engine, increasing bhp to 56. Lockheed disc brakes were introduced on the front wheels. The model was designated H/AN7 with chassis numbers beginning at 24732.

In the spring of 1964, a Mark III version (H/AN8), was introduced, chassis numbers beginning at 38829. The original quarter-elliptics were discarded and replaced by more conventional semi-elliptics. Winding windows and a new facia were featured. For 1967 came the Mark IV (H/AN9) but with a 1275cc (70 × 81mm) engine. Chassis numbers began at 64756. Production ceased in 1971.

Check points

Bodywork

On a somewhat smaller scale, the Sprite suffers from similar front-end problems to the E type Jaguar, in both instances the distinctive front ends presenting shortcomings of their own. On the Mark I cars, particularly, make a point of examining the seams for cracks and rust damage. Corrosion is also likely where the bonnet curves under at the front. Here the lip of the wing tends to attract road dirt on either side of the seam. In view of this, and various flexing problems, many cars are fitted with glassfibre fronts so an original steel one is well worth retaining as originality is very important. The

The Mark II Sprite of 1961 with re-styled front end and opening boot, all for £641

is hinged at the bulkhead and on opening it, check the state of the inner mudguards, particularly along their seams. Battery acid can cause rusting on the bulkhead and you may also find corrosion under the rain channelling at the top. An area where you are likely to experience real problems are the sills. They are particularly liable to rust at the front and rear and where the floor joins them inside the car. On early cars winding windows were considered unnecessary so you are unlikely to experience rusting problems on the doors, and side screens present few difficulties. The same can't be said for the door-mounting pillars on all of the cars, which can be found adjacent to the front bulkhead. Check the bottom hinge for signs of rot. Perhaps the most vulnerable area for rusting on the entire car is the rear box sections containing the quarter-elliptic back springs (Mark I/II). Here a persistent probe with a screwdriver has much to commend it and, if possible, also at the back of the seats where the strengthening plate for the rear suspension is bolted on to the floor. In addition, the inner rear wheel arch can rot placing a strain on the floor area, which *can* collapse as a result. The absence of a boot lid on the Mark I contributes to a fairly rust-free rear end, though a lid was fitted on the Mark II onwards.

Suspension, steering and brakes

The Sprite front suspension, borrowed from the A35, shares its strengths and weaknesses which are detailed on page 109.

The rear suspension isn't prone to problems, with the exception of the spring mountings already mentioned. However, it is advisable to check the state of the trailing arms for rusting, while, on Mark I and II cars, the bushing is suspect.

The rack-and-pinion steering isn't likely to cause many complications and the same goes for the Morris Minor-type brakes which are Lockheed hydraulics.

Engine, gearbox and transmission

The engine is the BMC A series power unit and was

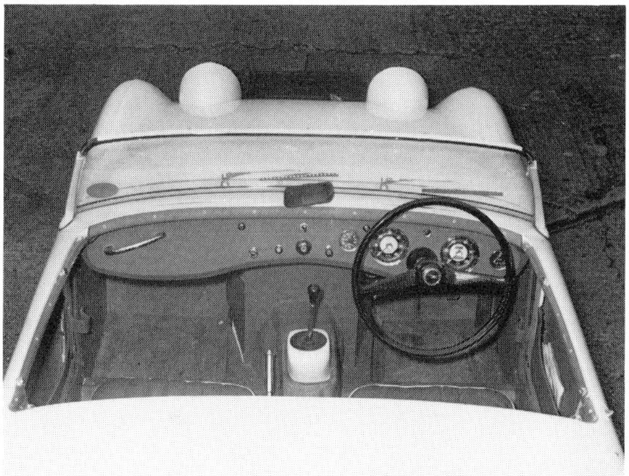

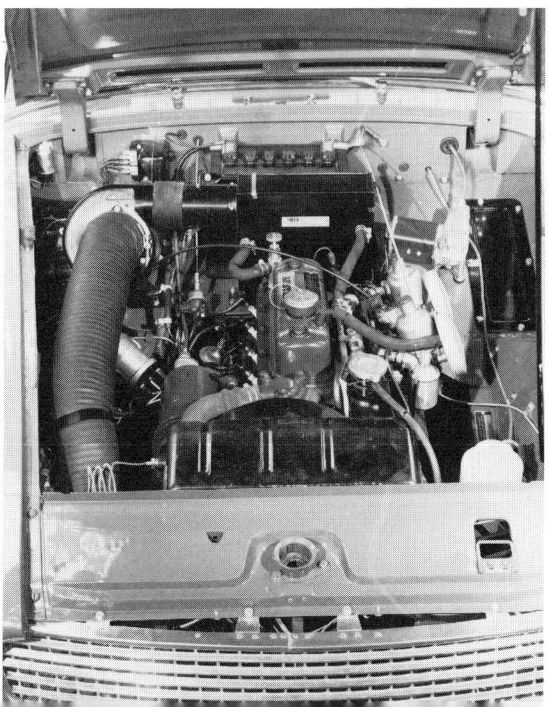

originally used in the A35. Its merits are set down on page 109. The same applies to the gearbox, though an inheritance is a noisy first gear and weak synchromesh on second gear. The rear axle, which isn't prone to break down, shares a differential with the contemporary Morris Minor, though not its half shafts; these are 2 inches longer!

Spares

Engine parts for the Sprite are sometimes available from the usual retail outlets, which is a major advantage; body panels are still a problem, though A. H. Spares of Leamington Spa are a useful parts source. See details on page 119. At present the Mark I 'Frogeye' bonnet assembly isn't available but the situation may change as the car becomes more of a collector's model. The Austin Healey Club has an active Sprite section; centre secretaries are listed on page 119.

Sprite performance and consumption (Mark II), *The Motor*, July 12, 1961.
Top speed: 82.1mph. Overall mpg: 38.6 for 2204 miles.

Austin Healey Sprite production

Mark I (HAN5)		48,999
Mark II (HAN6) 948cc		20,450
Mark II (HAN7) 1098cc		11,215
Mark III (HAN8)		25,905
Mark IV (HAN9) Abingdon built		13,861
Mark IV (HAN9) Cowley built		659
Mark IV (HAN9)		7421
Mark IV (AAN10)		1022
	TOTAL	129,532

Top, The Mark I's rather spartan cockpit. The rev counter was an optional extra. _Middle_, Mark I engine. Accessibility was good, a layout reminiscent of Jensen. _Bottom_, Mark II engine; a more conventional approach

A Mark VI Bentley somewhat incongruously pictured in a Waterloo Road car park with London bomb damage still evident in 1949

Walter Owen Bentley (1888–1971) founded Bentley Motors in 1919 but his first model, a 3–litre, did not go into production until 1921. This was succeded in 1926 by a more sophisticated 6½–litre 6–cylinder car, while the 4½–litre of the following year was a return to the 4–cylinder theme. Some 4½–litre cars were supercharged at the instigation of Bentley driver Tim Birkin. The mighty 8–litre six of 1931 was a carriage in the Rolls–Royce manner, but the Wall Street crash of 1929 had destroyed the market for such vehicles. A smaller 4–litre model was tried but a receiver was appointed in 1931 and production ceased after about 3000 cars had been built.

These are the bare facts of the Bentley story but there can be few vehicles more symbolic of the '20s (though a Morris Cowley is more typical) than the big green cars from London's Cricklewood. Bentleys were slightly old-fashioned in design and all of them (with the exception of the supercharged 4½–litre) were strong, reliable cars. Apart from the 4–litre, with which W.O. would have nothing to do, all were powered by sturdy engines with fixed cylinder heads and overhead camshafts. This reliability, which was a tenet of Bentley's design philosophy, paid dividends at the Le Mans 24–hour race, where Bentleys chalked up no less than five wins (1924, 1927–1930).

However, finances were always unstable and it wasn't until millionaire Woolf Barnato took over the company in 1926 that its survival was temporarily assured. Barnato spent £143,000 on Bentley Motors and in 1929 the company produced the only significant profit of the decade (£28,467). This figure dropped to barely over £1000 in 1930 and the following year a massive loss of £84,174 was recorded. It was in that year that Rolls–Royce was approached with a view to amalgamation but, after due consideration, the Derby company turned the idea down. Bentley went into liquidation but Napier, who had left the luxury-car market in 1925 to pursue their aero-engine interests, wanted to return and made an initial bid of £84,000 to purchase the company and its talented designer. This rose in stages to £103,675 and their final offer was a sealed bid of £104,775. However, the British Central Equitable Trust offered £125,175 and acquired control of Bentley Motors. The Trust was later revealed to be none other than Rolls–Royce who had successfully eliminated a potential competitor.

Production was transferred to the Rolls–Royce factory at Derby and in 1933 the 3½–litre 'Silent Sports Car' appeared, the engine being based on the RR20/25 unit. Later, in 1936, came the 4¼–litre model powered by a tuned version of the 25/30 and, like the earlier model, sported twin SU carburettors. Originally, the larger engine was to be offered as an option but it was immediately standardized and the model remained in production until 1939. A replacement car, the Mark V, was prepared for announcement at the 1939 Motor Show and some 20 or so cars were delivered up to 1940. This used the existing engine, fitted with a new chassis with cruciform centre bracing and independent front suspension. Also one experimental Corniche grand touring model was built, and therefore unique, it had a light chassis with bodywork by Van Vooren of Paris and was powered by an overhead inlet/side exhaust engine. It was dispatched to France for testing, in August 1939, but was destroyed by enemy action on the quay at Dieppe.

After the war, car production was transferred from Derby to Crewe. The first Bentley of this postwar era was the Mark VI, in essence the Mark V chassis fitted with the engine that had been used experimentally in the pre-war Corniche. The Mark VI Bentley was the first Rolls–Royce product to be offered with standardized bodywork, a package envisaged by RR's W. A. Robotham in his pre-war rationalization study, and detailed on page 218. The Mark VI benefited from a larger 4.5-litre engine in 1951, while 1953 saw the appearance of the R type, identifiable by its larger boot. Production ceased in 1955.

In 1952 a Bentley, far more in the marque's prewar traditions, appeared. The R type Continental had aerodynamic 2–door bodywork by H. J. Mulliner and was around 15mph faster than the standard R. At the time it was the fastest 4–seater saloon in the world with a top speed approaching 120mph. The series was discontinued in 1955. From thereon Continental versions of the S 1, 2 and 3 were produced.

With the arrival of the Rolls–Royce Silver Cloud and Bentley S Series, identities of the two marque's became completely fused. Details of these latter developments can therefore be found in the Rolls–Royce section on page 219.

Bentley S2 and S3 production

S2 (short)	1863
S2 (long)	57
Continental	296
S3 (short)	1286
S3 (long)	32
Continental	296

Mark VI 1946–1952
R type 1953–1955

Specifications and developmental history

The Bentley Mark VI was announced in May 1946. A 4–door saloon body by the Pressed Steel Company was mounted on a channel section chassis with cruciform centre bracing. Independent suspension was of unequal wishbone-and-coil-spring variety with Girling hydraulic brakes used on the front wheels. The rear ones were mechanical, both aided by a servo driven off the gearbox. Cam and roller steering gear was employed. The engine was a 4257cc (89 × 114 mm), 6–cylinder, overhead-inlet, side-exhaust unit. The interior was well appointed with leather-upholstered seats, walnut dashboard and door cappings, while the right-hand gearchange was an inheritance from pre-war days. The price was £2997.

A Rolls–Royce version of the design, the Silver Dawn, aimed specifically at the American market, was announced in July 1949. It did not become available on the home market until 1951. The Mark VI remained in production in basically unchanged condition until May 1951 when the engine's bore was increased from 89mm to 92mm, giving 4566cc. A full flow oil filter was fitted at the same time (from chassis B 2–MD).

The 1952 Motor Show saw the replacement of the Mark VI by the R type. It was substantially the same, apart from a larger boot, the car being lengthened by 7.5 inches and the rear suspension

Rolls–Royce equivalent of the Mark VI was the Silver Dawn which appeared for export only in 1949

strengthened to cope with the extra load. A minor change was the introduction of an automatic choke replacing the hand throttle and choke levers. Examples sold in 'dollar' markets were offered with an automatic option and this became available on the home market a year later. In April 1953 the engine's compression ratio was raised from 6.4:1 to 6.75:1 (chassis B93-TO), while in June 1954 the final drive ratio was altered to 3.42:1 from 3.727:1 (chassis B1-YA). Production ceased in 1955.

Check points

Body and chassis

The importance of getting an example in good structural condition cannot be over-emphasized. If you buy one riddled with rust you'll find yourself faced with a massive repair bill. First check points are the rear body mounts and it is by no means impossible that they have completely rusted away. These are to be found above the rear axle and you'll probably have to remove the rear wheels to examine them properly. While you're inspecting the rear of the car have a look at the inner body skin around the rear wheel arches. Rust can attack these arches externally and also the spats (if they're fitted) and there isn't much metal between the forward end of the arch and the rear edge of the back door. In addition, there are two more mounts at the back of the car near the spare wheel locker. The latter has a habit of attaining a lace-like appearance due to the ravages of rust, though fortunately the nearby boot floor usually escapes the worst of the onslaught. Open the doors, examine their lower edges and hinges as well as the bottom of the door pillars. After inspecting the state of the sills, move on to the front wings. First have a look at the side lights. These are leaded into the wing but mud can lodge in the knife-edge crease and it isn't long before rusting ensues.

A sunroof is an attractive fitting but it is, potentially another source of leakage. Open the hatch and check the condition of the channels. Water should drain away down tubes running through the windscreen pillars and if they get clogged up the channelling can suffer. If the locking pegs on either side of the roof rust then you might even have trouble opening it!

The chassis is very strong and at its strongest point is about 7 inches deep! It is not normally prone to rusting but it is worth checking the point at which it sweeps over the rear axle. At the same time examine the state of the car's exhaust system; replacements are expensive.

Suspension, steering and brakes

On the front suspension, the most likely point of wear is in the top and bottom wishbone pins and king pins. This is often caused by circulatory failure of the chassis lubrication system. Any oil leaks from this part of the car will pinpoint wear points. Rear suspension is by semi-elliptic springs and they are more likely to settle with age rather than break. The hydraulic shock absorbers are adjustable from the steering column and it is not uncommon for the adjuster to seize up through infrequent use.

The layout of the brakes is a development of those used on the pre-war Mark V Bentley, but the front shoes are Girling hydraulic units and the rear mechanically actuated by rods in tension. Perhaps the Lockheed master cylinder for the front shoes is surprising but the system works well enough. All drums benefit from a mechanical servo, driven off the gearbox, and this tends to operate reliably, except for the rear oil seal, which can leak.

Steering is by cam and roller and requires little attention.

Engine, gearbox and transmission

Early examples of the Mark VI were prone to seizing on long runs and a number of modifications introduced included a new variety of piston. Also the bypass oil filter fitted to these cars wasn't really suitable and, on occasions, bearing failures occurred. An increase in engine size, and a full flow filter in 1951, was a great improvement. An acceptable oil pressure for this engine is 20–25psi at 30mph (hot). Consumption should be around the 200-mile-to-the-pint mark.

The 4–speed gearbox is fitted with blocking plates on the top three ratios to prevent you from beating the synchromesh. A right-hand gear change should be a pleasure to use, but the distance pieces in the gearbox's layshaft have a habit of breaking up. You can get an idea of whether this has happened by warming the car up, selecting bottom gear and on a small throttle opening, feel for any kick-back in the lever. The box can also suffer from a noisy bottom gear but this can rumble on for ages without trouble.

Another wear point is in the centre bearing on the propeller shaft. The rear axle should provide years of reliable motoring.

A final word of warning. Make sure that if you buy a Silver Dawn (the Rolls–Royce equivalent) that it really is a Rolls-Royce and not a converted Bentley. Check the chassis plate and, if you have any doubts check with Rolls–Royce at Crewe.

The old magic, walnut dash and leather seats of a Mark VI Bentley. The lockable tool tray was a pleasing feature

Spares

Most mechanical parts for these cars are fortunately available from Appleyard Rippon of Leeds, who are the world-wide concessionaires for pre-1955 Rolls–Royce and Bentley parts. A number of firms will undertake restoration work on these cars. Just two are A and S Engineering of Farnham, Surrey and Michael Walker's Garage of Woking, Surrey. Both the Bentley Drivers' Club and the Rolls–Royce Enthusiasts' Club cater for these cars and are well-run, very professional organizations.

Bentley Mark VI and Rolls–Royce Silver Dawn spares and services

A and S Engineering, rear of 82/83 East Street, Farnham, Surrey (tel: Farnham 726488).
Appleyard Rippon Ltd., Roseville Road, Leeds 8 (tel: Leeds 32721).
Michael Walker's Garage, 5–7 Goldsworth Industrial Estate, Woking, Surrey (tel: Woking 62825).

Bentley Drivers' Club: Mrs B. M. Fell, W. O. Bentley Memorial Building, 15 Chearsley Road, Long Crendon, Aylesbury, Buckinghamshire (tel: Long Crendon 208233); Rolls–Royce Enthusiasts' Club. Lieut. Col. E. B. Barrass, Lincroft, Montacute Road, Tunbridge Wells, Kent (tel: Tunbridge Wells 26072).

The Mark VI's $4\frac{1}{4}$ litre engine. It was increased to $4\frac{1}{2}$ litres in 1951

Performance and consumption Bentley Mark VI (4.5–litre), *The Motor*, October 10, 1951.
Top speed: 100mph. Overall mpg: 16.5 for 174 miles.

Bentley Mark VI and R type and Rolls–Royce Silver Dawn production

Mark VI	4946
R type	2320
R type Continental	207
Silver Dawn	760

Bristol 1947 to present date

The Bristol Aeroplane Company's Car Division was founded in 1945 and right from the start the intention was to produce cars in small numbers to the high engineering standards demanded by the aircraft industry. The cars that went on sale in 1947 were based on the German BMW designs of the 1930s and the story of how this occurred is one of the most intriguing of the postwar years.

In 1934, H. J. Aldington, who manufactured Frazer Nash cars, was in Germany for that year's Alpine Trial, at which BMWs won the 1500cc team prize with Frazer Nashes taking second place. Afterwards, Aldington visited the BMW works in Munich and talked with the company's chairman, Franzt Popp. The outcome was the British sales and manufacturing rights of BMW cars, attained in the face of competition from, of all people, Austin. At the end of 1934, therefore, Aldington announced that the car would be available in Britain under the name Frazer Nash–BMW. Aldington, therefore, began importing the German cars, of which the 328 2–seater of 1936 is the most famous. Outwardly this car looked impressive being cleanly styled with faired-in headlamps. Under the bonnet the prospect was equally exciting. The engine was a 1971cc (66 × 96mm) six, based on the block of that used in the 320, 321, 326 and some 327 models. But Rudolf Schleicher, a BMW engineer, had converted this conventional pushrod layout, transforming it into a much more potent unit. A new aluminium head with hemispherical combustion chamber was fitted, but the really ingenious feature lay in the operation of the valve gear. Inlet valves, positioned over the side camshaft, were, conventionally, pushrod operated, but the exhaust ones opposite were actuated by rockers and short pushrods running *across* the head, all from one camshaft. The inlet tract was situated between the two lines of valves and breathing was excellent. The 328 soon built up an impressive competition career, achieving a fifth place at Le Mans in 1939 and special Touring bodied versions won first and third places in the shortened 1940 Mille Miglia. The war finally put a stop to these events and it wasn't until the end of hostilities, in July 1945, that H. J. Aldington, by then a Lieut. Col. in the REME, sought out BMW's bombed Munich factory and returned to England with the 328 which probably scored third place in the wartime Mille Miglia driven by Brudes and Roese.

Aldington was keen to transfer members of the talented design team to England to produce the postwar Frazer Nash–BMW. He had in mind designers Fritz Fiedler and Rudolf Schleicher (of 328 engine fame), Ernest Loof, a tuning expert, Von Rucker and Alex von Falkenhausen, an organizational genius.

Through his brother Donald, who had established contact with the Bristol directors while he was at the Ministry of Supply, Aldington opened negotiations with the aeroplane company, where he was actively supported by George White, son of one of the founders. They agreed to produce the car which was to be called the Frazer Nash–Bristol. Bristols purchased the manufacturing rights of both Frazer Nash and Frazer Nash–BMW cars from Aldington but H. J. would be able to assemble sports cars at his Isleworth factory under the Frazer Nash name, using parts supplied by the aircraft company.

This, at least, was the plan. First there were protests from the Americans, who were against the removal of materials and talent from BMW and then a section of the British motor industry protested at the importation of German brains and expertise. Consequently, Aldington had to settle for Fiedler, who was in prison at the time. After his release in 1946 he immediately went to Bristol to start work.

Little time was wasted and fortunately Aldington had obtained the blueprints and drawings of the BMW326 and 328 cars. The new Frazer Nash–Bristol was revealed to the motoring press in September 1946. It made its public debut at the Geneva Motor Show early in 1947, though by then it was known solely as the Bristol 400. The agreement was terminated at the aircraft company's request, the Frazer Nash name reverting to Aldington, though the supply of parts and skill to Isleworth was maintained. Dr Feidler later went to Frazer Nash where he designed the High Speed and Fast Tourer Frazer–Nashes of the postwar years.

The new Bristol was, in effect, the BMW326 chassis, 328 engine and 327-derived body. It was a close-coupled, 2–door saloon mounted on a robust, but not unduly heavy, box-section chassis. The subframe mounted front suspension, was a transverse spring with forged upper wishbone. At the rear was a semi-floating axle with arms, moving

George White, head of the Bristol Aeroplane Company's car division, complete with 401 scale model, in 1952. *Below*, the Touring bodied 328 BMW built for the 1940 Mille Miglia but with Frazer Nash grille. It provided the styling inspiration for the XK 120 Jaguar and MGA, amongst others

with it, connected to longitudinal torsion bars running along the frame and also to double-acting hydraulic dampers. At the axle's centre was an upper A-bracket to check any sideways movement. Rather surprisingly this chassis had not been designed by BMW in the '30s, but by the American Budd Corporation. Hydraulic brakes were fitted all round and rack-and-pinion steering employed. The engine, initially fitted with triple SU carburettors (though Solex soon replaced them) was the aforementioned 328 1971cc six improved by aircraft-quality materials. British thread sizes were also fitted. Inside the car the seats were trimmed in leather which set the seal of quality on the product.

The cars were hand-built, in the spirit of the pre-war Rolls–Royces, and an average of 150 cars a year was produced during the first 18 years. Bristol built practically everything themselves; only items like the tyres, wheels and carburettors were bought in.

Production of the 400 model started building up in 1947 and early the following year a few minor changes were made to the design. A conventional strip bumper replaced the original tubular one, while the boot was enlarged and the spare wheel mounted externally.

In October 1948 the 401 saloon and 402 drophead coupé appeared. Mechanically similar to the earlier model, the saloon was an aerodynamic *tour de force*. The shape had been created by Touring of Milan but greatly refined by Dudley Hobbs, after much experimentation in the Bristol wind tunnel. The result was one of the most efficient aerodynamic saloons of all time. The bodywork was constructed on Touring's Superleggera principles and was roomier than the 400. Distinctive features were the press-button, spring-loaded door locks and the sturdy alloy body-hugging bumpers with hard rubber inserts. However, these were replaced by chromium-plated metal strips with small over-riders on the 1951 model, the side scuttle vents also being discarded at the same time.

The potent, and beautifully-engineered Bristol engine, soon found its way into cars specifically designed for the race track. In addition to the Frazer Nash it was also used by A.C., Cooper, Lister and Tojeiro and one was fitted to the G type ERA. Bristols took over this last named project, changing it, to create the Bristol 450, with an unusual 2–door coupé with twin dorsal tail fins. A racing programme that embraced the Le Mans 24-hour race got off to a shakey start in 1953 when the experimental crankshaft gave trouble, but in the following 2 years, the Bristol team came seventh, eighth and ninth and won the 2–litre class.

In 1953 came the 403 saloon. This had the same body as the 401, which it replaced. A silvered grille and 403 bonnet flashes distinguished it from its predecessor. Under the bonnet was a 100bhp version of the 1971cc six in 100A form with a larger-diameter crankshaft and balance weights. A new camshaft and cylinder head with larger inlet valves rounded off the package. Additions to the suspension included a front anti-roll bar while the brakes were fitted with Al-fin drums. There was easier filling for the 17-gallon petrol tank and the rack-and-pinion steering gear was adjustable. Changes were few during the 403's production life though the 105bhp engine, used in the 404, appeared for the 1955 season.

The 404, announced in September 1953, was a luxurious short-chassis model having a radiator intake based on the wing leading edge of the Bristol Brabazon airliner. The last traces of the BMW radiator grille, a feature of the marque since the 400, disappeared with this model, while the other end of the car sprouted vestigial fins, echoing the racing 450. The spare wheel was mounted *inside* the nearside front wing as there was no exterior access to the boot. Rather surprisingly the aluminium body was clad on a pitch pine frame, though the doors used an alloy sub-structure. Two engine options were available: the 105bhp 100B or the hotter 100C variant, the latter being fitted to the Formula 2 racing cars of the day! In addition to this production model a 'cooking' version of the 404 was marketed in America by Chicago business man S. H. Arnolt. A 2–seater body by Bertone (Arnolt was vice president of the Italian coachbuilding firm) was fitted and the resultant package was sold as the Arnolt–Bristol.

The only 4–door Bristol appeared for 1955. This was the 405 and wood was again used in the construction of the body, though mostly above the waistline. The mechanics were largely unchanged, though a Laycock de Normanville overdrive was fitted and Michelin X tyres standardized. And it did have an opening boot!

There was a corporate shake up in 1956 with Bristols being divided up into three separate sections: Bristol Aircraft, Aero-Engines and Cars. Each were wholly owned by the Bristol Aeroplane Company and George White, the leading figure on the automobile division, became chairman and managing director of that section.

H. J. Aldington (right) who brought the Mille Miglia BMW to England after the war, with Count Lurani and Bristol 400 at Filton

The 2–door theme was re-introduced with the 406 model, the last to be powered by the BMW-derived engine. This was stretched to 2216cc (68 × 99mm) developing 105bhp. Dunlop servo-assisted disc brakes were fitted all round and the rear axle's location was altered to a transverse Watts linkage pivoting on a central casting and anchored to the chassis by mounting plates. The 406 went into production in 1958. The following year a mesh radiator grille replaced the vertically-slatted styling with fog lamps positioned within the radiator intake.

A further rationalization of the Bristol Aeroplane Company's interests came in 1959 when Bristol Aero-Engines merged with Armstrong Siddeley Motors to form Bristol Siddeley Engines and, in 1960, the facilities of Bristol Aircraft were absorbed into the British Aircraft Corporation. Bristol Cars Ltd., therefore, emerged as a private company with Sir George (having succeeded to the baronetcy) continuing as chairman and managing director and Anthony Crook, whose association with the marque went back to 400 days, becoming a director. It was an association that was to develop into a partnership in 1966.

Although the 407 of 1962 did not look dissimilar to the 406 under the bonnet was a 5130cc (98 × 84mm) V8 by Chrysler of Canada developing 250bhp. However, the engines were carefully rebuilt with high-lift camshaft and mechanical tappets in place of the hydraulic ones to permit higher revving. Torqueflite automatic transmission was employed. At last, the original transverse leaf-spring, independent front suspension was replaced by a coil-spring/wishbone layout. Although the familiar two-spoke safety steering wheel was retained (a feature of every model from the 401 to the 410 of 1968), the rack and pinion layout was changed for a Marles worm-gear layout. The slightly taller and narrower 408 appeared in October 1963. Mechanically, it was virtually the same as the previous model, but the twin headlights set in the radiator grille are a distinguishing feature. Electrically-controlled Armstrong Selectaride rear dampers were introduced. The 409 of 1965 had a 5.2-litre (99 × 84mm) version of the V8. Track was wider at the front (4ft 6in), while Girling servo-assisted discs replaced the Dunlop brakes.

400	**1947–1951**
401	
402	**1949–1953**
403	
404	**1953–1955**
405	**1955–1958**
406	**1958–1961**

(for specification and development history see previous text)

Check points

Body and chassis

Bristol bodies were mostly all aluminium which diminishes considerably the problems of corrosion, though this can be found in the area of the headlamps. It's also worth noting that the tops of the front wings are thicker than anywhere else, being 16– rather than 18–gauge. This prevented denting during assembly. An area where you can expect to find rust is on the steel rear wings on the 405/406 models, which tends to break out on the portion of the wing that serves as the door shut face. Also on the 401, the door bottoms were reinforced with steel and these can rust under their upholstery coverings. Similarly, all models that use a spare wheel well suffer from rusting on the steel sides of the box. The main snag with the 404 and 405 models is that they are ash-framed and this can rot over the years; replacement is a costly and time-consuming business. Look for evidence of corrosion around the boot access, windscreen surround, door A post and the cantrails in the roof.

A massive box-section chassis is used on all the Bristol models under review and it is most unlikely that you will experience any difficulties in that quarter. However, there are chassis extensions, front and rear, and these are prone to rusting. The front bumper is attached to the forward extension while the rear one plays a structural role in the design of the boot. At the front of the car the pan positioned between the arms can also rust. Fortunately, the Enots *One Shot* lubrication system usually manages to keep the front of the car in a well-preserved state as it can never be completely oil tight. One point to remember with a Bristol is

Left and right, the Bristol 401. Original styling was by Touring of Milan but the shape was enormously refined and improved at Filton. Construction was to Touring's Superleggera principles but again improved by Bristol engineers. Note the push button door locks and alloy bumpers with rubber inserts. *Below*, the 1961 406, the last Bristol to employ the BMW derived engine, in this instance of 2.2 litre. It cost £4244 new

not to attach a tow bar to that vulnerable rear chassis extension — you may well pull it right off!

Suspension, steering and brakes

A very straightforward front suspension system is employed. It is mounted on a detachable chassis member and uses a transverse leaf spring in conjunction with upper wishbones. The design changed little over the years, though the 400 and early 401s have the front shock absorbers integral with the wishbone pivot allowing easier access to the wishbone bearings, which is an advantage. The suspension is particularly durable as it is fed by the aforementioned Enots system. At the rear, longitudinal torsion bars are used and the trunnion bearings at the point where the damper arms join the rear axle are lubricated by the axle's oil. If the seals wear, lubricant will escape so take a close look for any signs of seepage.

A proprietary rack-and-pinion steering unit with purpose-built ball joints is used, the latter being automatically lubricated. There should be no play in the rack and if there is the pinion is mounted in an eccentric bush, which can be adjusted to take up slack on most models.

The brakes on most models are admirable and reasonably trouble free, though perhaps the 401 requires more pedal pressure than most.

Engine, gearbox and rear axle

The 2–litre, 6–cylinder engine was much the same up until the 405 model, the 406 having a bigger bore to increase the capacity to 2.2–litres. Fortunately, it is a very reliable unit and a good oil pressure reading is 60psi at 3000rpm at 70 degrees. Anything below 40psi is an indication of impending trouble. This is a crucial point if you're considering buying a Bristol in need of an engine overhaul, because at the time of writing (1980) you won't have much change from £2000 for a complete recondition and, though parts are available, they are expensive. Having said that, once the engine has been overhauled it should not require any further attention until it has exceeded well over 100,000 miles. One problem which affects the engine is the crankshaft damper. The difficulty can be identified by a rather unnerving whirring emanating from the lower reaches of the engine after switching off. The only other shortcoming occurs on the 406 model, which tends to run hotter than other examples, so the state of the radiator and associated pipes is, therefore, more crucial.

If you decide to become a Bristol owner and the engine's aluminium cylinder head has to be removed for any reason, it must be lapped in to the block, prior to replacement. Nowadays a composite cylinder head gasket is used in place of the original

steel shim one which demanded somewhat closer tolerances, but the job is still a very necessary one. Its also worth noting that Bristol engines are fitted with tiny 10mm sparking plugs. Fortunately, the Japanese use this size for their rotary-engined cars and also their motor cycles and Bristol Owners' Club members find that NKG C6M or C7HS are suitable in most cases. The valve adjusting screw has a 40tpi thread, which incidentally, is a Model Engineering size.

The exhaust system runs from two separate manifolds via pipes to a single silencer. In the best Bristol traditions, the exhaust pipe locking rings are made of phosphor bronze and the threads must be carefully lapped in with grinding paste before assembly. Burr marks on the rings will reveal if the car under inspection has been in the hands of an unsympathetic owner!

The gearbox uses an aluminium casting split vertically, just like a $6\frac{1}{2}$– or 8–litre vintage Bentley. It is beautifully made, as might be expected, Bristol initially manufacturing it, though later Borg–Warner took over. The only major problem you're likely to experience is with first gear. This incorporates a free wheel unit and rapid acceleration in first cog can result in broken rollers. At best this means the loss of one gear, though at the other extreme the whole box can seize. If you're faced with a Bristol with a damaged box be prepared for a big repair bill, a replacement first gear alone is *very* expensive.

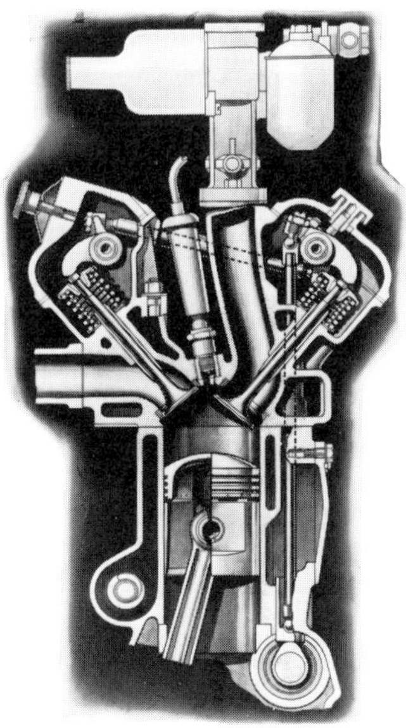

Above, the Bristol's engine is easier to illustrate than describe! The side camshaft and cross pushrod layout are clearly shown, together with the 10mm sparking plugs. Solex carburettors soon replaced the SUs shown. *Left*, the 401's engine. Breathing was good but a rather tall unit resulted from the ingenious layout

Rear axles aren't subject to any particular problems.

Spares

Fortunately many body, as well as mechanical components are still available from Bristol Cars which is a major advantage. However, difficult parts to obtain on the 6–cylinder cars are engine blocks, cylinder heads and water pumps. Exterior trim is difficult to find and it is worth noting that the Standard Vanguard employed the same exterior door handles as Bristols from types 404 to 409. There are now no new carburettor bodies available and triple Solex B32s were used, though the 3–litre two-stroke Saab used a similar unit, which is worth noting. Throttle chambers and jets are still available. TT Workshops of Westbury, Wilts and Windley Restorations of Wokingham, Berks will undertake work of these finely engineered power units. The enthusiastic Bristol Owners' Club will also be pleased to hear from you.

Bristol spares and services

Bristol Cars Services Ltd., Great West Road, Chiswick Flyover, London W4 5QW (tel: 01-994 3417/8/9).
TT Workshops Ltd., 127, Engineers Road, West Wilts Trading Estate, Westbury, Wiltshire (tel: Westbury 3603).
Windley Restorations Ltd., Denewood, Coleshill Farm, Nine Mile Ride, Wokingham, Berkshire (tel: Eversley 732088).
Bristol Owners' Club, Bernard Lee, 2, Middle Green Road, Langley, Slough, Berks.

Performance and consumption Bristol (400), *The Motor*, May 19, 1948.
Top speed: 94.1mph. Overall mpg: 21.4 for 1009 miles.

Bristol production 1947–1961

400	700
401	650
402	24
403	300
404	40
Arnolt	142
405	297
405 Abbott	43
406	300
406 Zagato	6

Daimler 1896 to present date

One of the oldest British car concerns, the Daimler Motor Company, was created in 1896 by Harry J. Lawson who had high hopes of controlling the embryo motor industry. A disused cotton mill was acquired in Coventry for the production of Daimler cars, though output did not begin until the following year. Prior to that the company had relied on screwing Lawson's patent plate to Panhards and Cannstatt-Daimlers! Although the first model was a 4HP twin, the number of types increased alarmingly. (The Prince of Wales purchasing a 6HP Phaeton in 1900.) However, the range of models was pruned in mid-1902 and just three types were available. The company enjoyed Royal patronage again, in 1902, when King Edward VII (as he had become) purchased a 22HP model, forging a connection with the make that was to endure until the 1950s.

Although much of Lawson's shambling empire collapsed in the early years of the century, Daimler survived the upheavals and the company was reformed in 1904 when the distinctive finned, though anonymous, radiator grille appeared. In 1909 Daimler took up Charles Yale Knight's double-sleeve valve engine in a bid for greater silence and the following year the company merged with the Birmingham Small Arms group. Consequently the Daimler and BSA ranges were more rationalized.

The glass fibre bodied Daimler SP250. Announced in 1959, it remained available until 1964

These well-mannered and stately vehicles whispered their way through the '20s with an extraordinary variety of models; in 1927, for example, no less than 27 were listed! In that year the 7136cc Double Six appeared, having a V12 sleeve valve engine designed by Laurence Pomeroy, the company's chief engineer.

In 1930 the company pioneered the fluid flywheel in conjunction with the pre-selector gearbox that became a feature of all Daimlers from 1932 to 1956. Lanchester was acquired in 1931 and before long sleeve valves were on the way out, conventional poppet valves appearing on the Pomeroy-inspired smaller, 6–cylinder 15HP model of 1933. Sleeve valves lasted only another couple of years; by 1936 poppets reigned supreme. In 1938 the 15 acquired independent front suspension and was the most popular of a range that embraced three sixes and two straight eight models.

In 1937 the company departed from the original Motor Mills site and production was concentrated at nearby Radford. (The original Daimler factory was destined to be a victim of the Coventry blitz.) Daimler, who had been at Radford since 1908, gradually extended their works and these were further extended by the establishment of a shadow factory there in 1937. Bristol Pegasus aero-engines were manufactured and later shadow factory Number 2 was built at Allesley.

During the war, scout and armoured cars were produced by Daimler while Bren and Browning gun parts were made in large numbers. However, they still found time to consolidate their motor manufacturing interests when, in 1940, they purchased the well-established coachbuilding firm of Hooper and Co., after many years of association. Hooper had, in 1938, bought up Barker, their old rivals; thus two illustrious names came within the Daimler orbit.

After the war the company pruned its range, though the largest model, the 5.5–litre Straight Eight, which remained available until 1953, was a pre-war anachronism. It is probably best remembered as forming the basis of the Docker Daimlers that so enlivened postwar motor shows between 1951 and 1955. Later examples used a 4.5–litre chassis.

The two other models of the immediate postwar era were sixes, the 4.1–litre DE27 and the 2.5–litre DB18, the former remaining in production until 1951. The DB18 was revised as the Consort for 1950, with a curved radiator grille and faired in headlamps. A drophead coupé version of the DB18 was replaced by the Barker Special Sports Coupé for 1949 having a bench-type front seat and transverse rear one. Both this and the Consort were dropped in 1953 and replaced by the Conquest (Series DJ) saloon. This was powered by a new 2.4–litre six with a laminated torsion-bar-and-wishbone, independent front suspension system, instead of the coil-and-wishbone layout, used on other Daimler models. It was soon joined by coupé and roadster versions. In 1954 came the more powerful Conquest Century, distinguished by a chromium-plated windscreen surround and rear window, the lower-powered Conquest model being dropped in 1956. For 1957 the Conquest Century was offered with the option of Borg–Warner automatic transmission.

Daimler's contribution to the large car sector of the market was re-established with the DK400 of 1955. This used a 127 bhp 4.6–litre six and remained in production for 2 years. There was also the Regency II (the Mark I had been a 3–litre that never went beyond the prototype stage in 1951) and Sportsman. They were available in 3.5– and 4.5–litre forms and, superficially, resembled the smaller-engined Conquest series. This rather unwieldy range of models does, nevertheless, reflect a certain lack of direction in the company's affairs; in 1956, for instance, no less than 10 models were offered.

For 1957 the DK400 engine was increased in power and the bodywork standardised in A or B de luxe form. 1958 saw the appearance of the 3.8 litre Majestic saloon with automatic transmission, while the following year there was a complete break with Daimler traditions with the announcement of a glass fibre bodies sports car, the SP250. Under the bonnet was a 2.5 litre V8 engine, designed by Edward Turner and a larger version of this power unit appeared in 1960, this time in 4.5 litre form. It was fitted in a revised version of the Majestic saloon, re-named the Majestic Major though the straight six continued in production until 1962. A 4.5 litre V8 limousine was also offered.

In 1960 Jaguar purchased the Daimler company and the first result of the union appeared for the 1963 season. This was the 2.5–litre V8 250, the smaller V8 being transplanted to the Jaguar Mark II bodyshell and was initially only available with automatic transmission. In 1966 the new Daimler Sovereign was announced, though this was a badge-engineered version of the Jaguar 420 saloon, complete with XK engine. Both Daimler V8s were out of production by 1969; thereafter the marque became a badge-engineered, but up-market, variation on the Jaguar theme.

Known Daimler production 1946–1967

DB18 $2\frac{1}{2}$–litre saloon and foursome coupé	2503*
DB18 Consort saloon	4250*
DB18 Special Sports Empress saloons and other Hooper specials	108
Regency 3 litre saloon (Series DF300)	12
3–litre Hooper bodied Empress saloon (Series DF302)	32
3–litre coupé (Series DF302)	4
$3\frac{1}{2}$–litre coupé (Series DF304)	395*
Sportsman and Empress saloon and one or two specials (Series 306–9)	58*
$4\frac{1}{2}$–litre Regency Mark II and Sportsman (Series DF400)	28*
One-O-Four and Lady's Model (Series DF301)	559*
DK400 custom built	16
DK400 standard Limousine	106*
DE27 (all coachbuilt)	205
DH27 Hooper limousine for Daimler Hire Ltd	50
Straight Eight (Series DE36), all coachbuilt	216*
$2\frac{1}{2}$–litre and V8 250 (to 1969)	17,915
Sovereign (to 1969)	5850

*indicates estimated figure

SP250 1959–1964

Specifications and developmental history

The Daimler Dart was announced in April 1959 and launched at the International Automobile Exhibition in New York. After protests from Chrysler, who owned 'Dart' for their Dodge marque, the name was changed to SP250. The 2–door 2–seater bodywork was of glass-fibre construction, mounted on a box-section chassis. Independent front suspension was by coil-and-wishbone and Girling disc brakes were fitted all round. Cam Gears cam and lever steering was employed. The engine was a 2548cc (76 × 69mm) pushrod V8 developing 140bhp. Seats were upholstered in leather and the price was £1395.

In spring 1960 a new style facia and final drive ratio were introduced. The B Specification models appeared in April 1961. By this time additional steel box sections had been added to the sills, rear body and scuttle mountings for greater rigidity. The glass-fibre shell was also strengthened and those items, including bumpers, adjustable steering column, petrol reserve unit, screen washers and exhaust finishers, previously listed as optional extras were standardized.

C Specifications appeared in March 1963. A heater/demist unit was standardized along with a trickle charger socket and cigar lighter. Production ceased in January 1964.

Check points

Body and chassis

The model has glass-fibre bodywork, which is up to half an inch thick in places and extremely strong. Any problems are, therefore, confined to stress cracking, particularly around the headlamps, in the vicinity of the door handles, the doors themselves (individuals tend to lean on them) and the rear wings. Cracking can occur on the bonnet as it is fairly flexible and some have been known to jump their safety catches with disastrous results.

But real trouble can be experienced lower down! Areas that require particular scrutiny for rust are the tube holding the steering idler, the bottom of the front suspension mounting pillar, as it lacks a drain hole, and the front hanger of the rear spring. After the Jaguar take over of the company they added extra box sections between the chassis outriggers, in the form of mild steel sills, and up behind the doors. And these have a habit of rusting away to nothing.

Suspension, steering and brakes

The front suspension is practically the same as the TR3 with the exception of the vertical link and stub axle and coil spring. It is important that the link gets plenty of lubrication and if it hasn't been regularly attended to then this is your most likely wear point.

Steering is like the TR or Austin Healey 3000. The only difference is that the box is mounted upside-down, so oil may seep out and require more regular attention than normal. Ride height for the Daimler is 6.25 inches between the ground and the chassis and it's worth noting that, if the suspension starts to tire, the steering box will probably be the first item to be damaged.

The disc brakes shouldn't present any major problems, though the supply of rear calipers is diminishing. Brake pipes are prone to corrosion, particularly those on the rear axle together with the one running from the front of the car to the nearside rear. Also check the point at which the pipe passes through chassis members.

Engine, gearbox and transmission

The V8 engine gives remarkably little trouble apart from a tendency to consume oil, either by leakage or combustion. The latter symptom is caused mainly by wear of the valve guides, which also have a tendency to move slightly and cause a sound like noisy tappets. At the bottom of the engine, the centre main bearing can be troublesome. If you have any doubts on this score, drive the car at about 40mph in top gear and then accelerate rapidly. If the oil pressure gauge dips and recovers then bearing troubles may be imminent. Low pressure can also be caused by a worn oil pump; it is not uncommon for end float to build up in the gears. However, if the mains are worn, the centre three can be replaced with the engine *in situ*. The rear is tricky and, if the front requires attention, the engine will have to be removed. A slipping clutch isn't unusual with this model as it is reckoned that the 9.5–inch diameter unit wasn't really suitable. The gearbox, which is again based on the TR3, is usually considered to be the weakest part of the car with a clattering first gear and no synchromesh. The primary shaft, which isn't shared with the TR, is a difficult spare to come by.

Rear-axle spares are tricky, though it is worth noting that the specifications were improved with production — the later 7HA unit is the one to fit if your existing one gives trouble.

Spares

You may be lucky and obtain engine parts from retail outlets as the engine was used in the Daimler 250 saloon (Jaguar Mark II shape) until 1969. It is worth noting that the Lucas distributor for this V8 tends to be expensive. Two spares sources are David Manners of Birmingham and Barry Thorne of London and, of course, the Daimler and Lanchester Owners' Club as they have an active SP250 section and can also supply reprints of contemporary handbooks, spares lists and the like.

SP250 spares and services

David Manners, 17 Hagley Road West, Birmingham B17 8AL (tel: 021-429 1433); callers by appointment.
Barry Thorne, 57 Northcote Road, London SW11 1NP (tel: 01-228 6835).

The 250's cockpit, inevitably a 1959 export version

Under the bonnet. The water filler has been lowered to permit a low bonnet line

Daimler and Lanchester Owners' Club, Barry Thorne, address left.

Daimler SP250 production

Left-hand drive		1200
Right-hand drive		1445
	TOTAL	2645

Ford

1911 to present date

As already recounted, up until 1932 Ford of England largely produced right-hand-drive versions of the American Models T and A, the exception being the small bore AF, marketed in deference to the European horsepower laws. However, the new Dagenham factory really got into its stride with the 8HP Model Y of 1932. Production ceased in 1937, and it was replaced by a re-styled Eight that was made up until 1939. Slightly larger were the 1172cc C and CX models which appeared in 1934; those were available until 1937 when they were replaced with a re-styled Ten similar to the new Eight. The following year the Ten was re-designed to become the 4–door six light Prefect. Top of the range was the V8, introduced in America in 1932, but up until 1935 examples sold in Britain were produced in Canada. The V8 and its variants were destined to sell in small numbers as their large-bored engines were penalized by the British horsepower taxation laws. Therefore, the V8-62 of 1935–1941 used a smaller-capacity 2227cc engine instead of the more conventional 3622cc unit as fitted to the V8-48 of 1935, the V8-68 of the following year, the 78 of 1937, the 81 of 1938 and the 91 of 1939.

During the Second World War, Ford's mighty Dagenham works produced thousands of vehicles ranging from small vans to four-wheel-drive trucks and Fordson tractors. In addition a new factory was established at Urmston, Manchester for the manufacture of Rolls–Royce Merlin aero-engines.

Production re-started almost immediately after the war. Up until the 1950 Motor Show, when the first of a new range of cars was announced, this consisted exclusively of pre-war products. The 933cc Anglia (Series EO4A) had appeared in 1939 and this was made until 1948 when it was slightly modified, the running boards were eliminated and the radiator grille altered, continuing in this E494A form, until 1953. Similarly, the Prefect (E93A) of 1939 was slightly changed in appearance for the 1949 season, the more obvious alteration being headlamps grafted into the front wings (E493A), and this again was dropped in 1953. The largest capacity model of the range was the 3622cc V8 Pilot (E71A) of 1947–1951 but this was a short-lived low-production affair. All the models so far chronicled had side-valve engines fitted in channel section chassis and sporting transverse leaf suspension. By contrast the new Consul (EOTA) and Zephyr (EOTTA), both announced in 1950, were modern in every way, with overhead-valve oversquare engines mounted in full width monocoque bodywork and employing MacPherson strut independent front suspension. The smaller-engined Consul used a 1508cc 4–cylinder power unit while the six in the Zephyr was 2262cc. For 1954 came the Zodiac, a more powerful and better-finished version of the Zephyr and 2 years later a Mark II version of all three models appeared. The styling was updated and engine capacity increased in the Consul (204E) to 1703cc, while the Zephyr/Zodiac (206E) six was stretched to 2553cc. These all remained in production until 1962.

The new broom didn't reach the bottom end of the range until the 100E model was announced in 1953. This was initially available in 2– and 4–door saloon form and used an 1172cc engine, in the Prefect traditions. At the same time the archaic side-valvers were dropped, with the exception of the Popular (103E). This was basically the Anglia body fitted with the 1172cc Prefect engine, already featured on some export Anglias. It remained available until 1959 and, for a time, was the cheapest car on the British market. When it was phased out the Popular name was transferred to the 100E, as the Anglia was needed for a new model which was duly announced for the 1960 season. The 105E Anglia, with its distinctive inward-inclined rear window, followed an engine theme established with the larger cars; an oversquare overhead valve four, in this instance, of 997cc. This engine was also fitted in the 100E body and the resultant model of 1959–1961 designated the 107E Prefect. In 1961 came the Classic (109E) being the first disc-braked Ford with a 1340cc engine, though it was also available the following year in 1498cc (116E) form. The model, along with the 2–door coupé Capri, was dropped in 1964.

In 1962 came the best-selling Consul Cortina (113E), though the later version for 1965 was plain 'Cortina'. Initially only available in 1198cc form, it was joined for 1963 by the 1498cc Super version (118E). Both were produced in 2– and 4–door saloon and estate-car form. The Ford Motor Company's continuing involvement with motor sport was reflected by the Cortina Lotus (125E) of 1963, powered by a 1558cc twin-cam engine, as fitted to the contemporary Elan and assembled by Lotus. Later in the year came the disc-braked GT

A 1956 100E Anglia, pictured at Osterley Park, Middlesex. The horizontal radiator slats are an identifying feature

(118EGT), relying on a tuned version of the 1498cc engine. The Mark I Cortina was replaced for 1967 by the re-bodied Mark II and all versions were disc braked. A new 1297cc engine became available (3034E), with the 1500cc version (3036E) was inherited from the previous range. The Corsair (120E) of 1964 was initially available with the 1498cc Cortina engine, as was the GT version (120GT), though from 1966 V4 engines of 1663cc and 1996cc were fitted. Production was continued until 1970, the same year that the Mark II Cortina was phased out.

The Consul name had been transferred, in the first instance, to the Cortina so Ford's large cars carried only the Zephyr and Zodiac name in Mark III form. In 1962 the former model was offered with a 1703cc 4-cylinder engine (221E) and the 2553cc 6-cylinder one (213E). The Zodiac was available only with the 6-cylinder power unit. These were made up until 1965, with the Mark IV becoming available the following year. The Company's new Vee engines were used to power the range which offered, for the first time, independent suspension all round. The Zephyr was fitted with a choice of power units: the 1996cc (3010E) or 2495cc (3012E), though the Zodiac was only available with the 2994cc (3008E) engine. An Executive version, with the 3-litre V6 (3022E) went on sale for 1967. The range ceased production in 1971.

Mention should be made of the GT40 project which gave Ford their first Le Mans win in 1966, following intensive development on both sides of the Atlantic. Other victories at the classic 24-hour race came in 1967, 1968 and 1969. During the 1960s, Ford had certainly achieved considerable success in competitive events and excelled in car production.

Ford production 1945–1967

Anglia EO4A (from 1938)	50,684
Prefect (from 1938)	379,229
Mark I Consul	231,481
Mark I Zephyr	152,677
Mark I Zodiac	22,634
Popular 103E	155,350
Mark II Consul	350,000

The Squire of 1956 with its wooden side mouldings was the estate car version of the Prefect

Mark II Zephyr	221,000
Mark II Zodiac	80,000
Anglia 105E saloon	954,426
Anglia 105E estate	129,529
Classic and Capri	126,206
Mark III Zephyr (four)	105,236
Mark III Zephyr (six)	107,006
Mark III Zodiac	77,323
Mark I Cortina	1,010,090
Corsair	329,676
Mark IV Zephyr	41,389
Mark IV Zodiac	46,836
Mark IV Zephyr V6	61,022
Mark II Cortina	1,010,090

100E 1954–1962
107E 1960–1961

Specifications and developmental history

The Ford 100E was announced at the 1953 Motor Show. It was a sturdy monocoque available in 2-door (Anglia) and 4-door (Prefect) form. MacPherson strut independent front suspension was fitted while the steering was by worm and peg. Girling hydraulic drums brakes were employed. The 4-cylinder, 1172cc (62 × 92mm), side-valve engine was similar to that fitted to the earlier Prefect model, but was a completely new casting with larger bearing surfaces and inlet valves, a higher 7:1 compression ratio, adjustable tappets

The four door Prefect, with vertical radiator slatting

Ford's last side valver. The two door Popular was available until 1962

and a water pump. It developed 36bhp. The Anglia was upholstered in PVC and had a three-dial instrument panel. More stylish semi-circular speedometer was a feature of the Prefect, which was offered with a choice of PVC or leather upholstery. The Anglia, with its horizontal radiator slats, cost £511, while the Prefect, with vertical slats, sold for £560.

After the beginning of 1955 the brake-drum diameter was increased by an inch. Up until March the sump capacity was 5.25 pints, which included 0.75 pint for the filter. From April 1955, the contents of the sump were reduced to 4.5 pints, of which the filter accounted for 1 pint.

De Luxe versions of the two models appeared for the 1956 season. Identifying features are chrome waistline strips, while inside the trim was improved and a moulded plastic facia was introduced in place of an instrument binnacle. At the same time an estate car was introduced. The Escort was the Anglia version, while the Squire was the Prefect's equivalent and was distinguished by the addition of wood side-mouldings and a more luxurious interior. In 1959 all models were phased out, and replaced by the single Popular model, which was made up until 1962.

The 100E body style was perpetuated for the Prefect 107E, introduced for 1960. Mechanically it incorporated most of the features from the new 105E Anglia. The engine was a 977cc (80 × 48mm), overhead-valve, pushrod, 4–cylinder engine, developing 39bhp. A new 4–speed gearbox and hypoid rear axle were all part of the same package which sold for £621. Main identifying features for this model is an additional exterior chrome moulding, which ran from the front of the main chrome side strip to the front wheel arch, while inside improved seats were fitted along with pile carpets. The exterior modifications permitted two-tone paintwork which was adopted by most buyers. Production ceased in 1961.

Check points

Bodywork

The rear of the car is rather more vulnerable to rust than the front, so let's start there. Look around the rear wheel arch and, if you're contemplating buying a 4–door version, the part of the rear door that forms the rear wheel arch. Make a point of checking the condition of the boot as it is particularly prone to condensation and it is not uncommon for it to rust from the inside. On the 100E model, it is advisable to remove the rubber bung in the spare wheel well and drain out any water that may be present.

Moving towards the front, examine the two jacking points (there is one on each side), which can give trouble. Examine the sills at the same time; it is unlikely that they will be rust-free. Inspect the bottom of the front wings adjacent to doors, as this is another favourite rust point. Fortunately, the wings are bolt-on units and if you're lucky enough to find any steel replacements then fitting isn't too difficult.

Suspension, steering and brakes

The suspension checks are particularly vital as you can expect trouble of similar severity at opposite ends of the car! Starting at the back, carefully examine the state of the rear spring hangers, particularly the forward one — you may be in for a nasty shock. Then to the front; raise the bonnet and examine the tops of the MacPherson strut front suspension units, which you'll find on either side of the engine compartment. The area immediately around the mounting is liable to corrosion which is accelerated by stress. At the top of the strut is a bearing that also wears. To check its state, grasp the top of the road wheel and rock it backwards and forwards. If the top bearing is worn you can *see* the whole strut moving inside the wheel arch.

Brakes should give little trouble, but you'll need a hub puller to get the rear drums off. Incidentally, wheel cylinders are shared with the contemporary Triumph Herald.

Engine, gearbox and transmission

The virtually indestructible Ford side-valver will go on practically indefinitely, though it is vulnerable to main-bearing wear. These are of the old-fashioned white metal variety but, should the

The Prefect's interior. The semi-circular instrument panel distinguished the model

The Prefect's 1172cc side valve engine though the 12 volt battery appeared for the first time on a small Ford. *Right*, the Cortina Lotus of 1963. The green side flash and Lotus badge on the rear wing identify

engine require reconditioning, it can be converted to shells. Another shortcoming is a tendency for the top piston ring to break up and, if the car smokes badly, this is the most likely cause. Another top-end snag is a proneness on the part of the cylinder head to crack in the web immediately adjacent to the thermostat housing and, what might appear to be a dribble from the water hose could be something much more serious.

Unfortunately, the 100E's gearbox is a 3–speed unit and second gear suffers. So be prepared for wear in the gear itself and in the synchromesh. Transmission 'clunk' is usually caused by a worn rear universal joint; the heat from the rear axle dries out the grease in the joint. In fact, the car is really better off with a joint that can be regularly lubricated — a worthwhile modification. Rear axle is usually trouble free.

Spares

Mechanical spares for this model are still plentiful and you might even obtain them through the usual retail outlets. If not, try Shepherds Grove Service Station of Stanton, Suffolk or the Newford Parts Centre of Chorley, Lancs. Also the active and enthusiastic Ford Sidevalve Owners' Club is continually investigating the re-manufacture of parts, so if you have a 100E, it's well worth joining.

100E spares and services

Newford Parts Centre, Fishwick Lane, Higher Wheelton, Chorley, Lancashire (tel: 0254 830343).
Shepherds Grove Service Station, Stanton, Suffolk (tel: Stanton 50289 (day); Botesdale 668 (evenings)).
Ford Sidevalve Owners' Club, John Norris, 6, Thakeham Close, Bexhill on Sea, Sussex.

100E performance and consumption, *The Autocar*, February 26, 1954.
Top speed: 68.3mph. Overall mpg: 29.7 for 375 miles.

100E production

100E Anglia		348,841
100E Prefect		100,544
Escort		33,131
Squire		17,812
	TOTAL 100E	500,328
107E		38,154

Cortina Lotus 1963–1966

Specifications and developmental history

The Cortina Lotus, designated 125E by Ford and

type 28 by Lotus, was announced in January 1963. The bodywork was a Mark I Cortina 2–door saloon, finished in white with green side flashes. Aluminium doors, bonnet and boot lid were departures from standard. MacPherson strut independent front suspension with lower springs and stiffer dampers was fitted, while zero front wheel camber was achieved by employing slightly longer forged, rather than pressed, track control arms. A stiffer anti-roll bar was also fitted. The Cortina's rear semi-elliptic springs were discarded and replaced by trailing radius arm/A bracket system with coil springs. Brakes were Girling servo-assisted with front wheel discs. $5\frac{1}{2}$J wheels were fitted. The steering gear was of the re-circulating ball variety. The engine was a 1588cc (82 × 72mm), twin-overhead-camshaft, Ford-based unit, as fitted to the Lotus Elan, and developing 105bhp. The interior of the car was special with bucket front seats and a 15-inch wood-rimmed steering wheel. Price was £1100.

From July 1964 a two-piece propeller shaft replaced the one-piece unit. A racing model was announced the same year boasting a slightly larger 149bhp 1593cc engine. Models in 1965 benefited from the Ford aero-ventilation system, instantly identifiable by a small grille adjacent to the rear window. In September 1965 semi-elliptic rear springs replaced the coil-spring and A bracket arrangement, though five leaves, instead of the customary four, were fitted. Models for 1966 used the close ratio Corsair V4 gearbox. Production ceased in 1966.

Check points

Bodywork

The Cortina Lotus suffers from most of the body shortcomings of the conventional Mark I Cortina, and a few more besides! Your first check point should be the bodywork immediately around the headlights and, if rust is well established, it can corrode the lamp reflector as well. The bottom of the front wing immediately adjoining the door is another area worthy of investigation. Look underneath the car as the outrigger/jacking points are prone to rusting badly, particularly the nearside one. The rear wing valances and the state of the boot floor can produce some unpleasant surprises. On the nearside is a spare wheel well, though it doesn't contain one as the $5\frac{1}{2}$J-rim wheels are too large. Unfortunately, the well, as its name implies, is a water trap. On the opposite side of the boot is the battery. (On the basic Cortina this is placed under the bonnet but in the Lotus model, there isn't room in the engine compartment so it is relegated to the boot.) And we all know how metal around a battery can become corroded. While checking the boot, look at the shock absorber top mountings which project into the floor. Rust does have a habit of corroding the turret walls, so check these carefully.

The aluminium bonnet and boot lid doesn't rust but remember that the alloy is very thin and is therefore easily dented.

Suspension, steering and brakes

Most of the Cortina Lotus's suspension and steering is peculiar to the car. This applies to the MacPherson struts, track control arms, anti-roll bar and bushes, steering drag link and steering arm and ball joint. The same applies to the A bracket rear suspension, which we will discuss presently. Check first the MacPherson strut mountings, which are situated on each side of the engine compartment under the bonnet. These mountings are subjected to a great deal of stress and are also bombarded with mud thrown up by the wheels. Watch for signs of bodged repairs and remember that plates welded on to reinforce the area is not a guarantee that all is well underneath. . . . For the first 2 years of production the Cortina Lotus was fitted with a Chapman-designed rear suspension with coil springs replacing the conventional semi-elliptics and the axle located by an A bracket and radius arms. This was an area destined for radical alteration because the loading on the aluminium differential housing resulted in excessive oil leakage. The lubricant found its way on to the rubbers of the A bracket mounting triggering a chain of events that led to the collapse of the system. From late 1965 the cars reverted to a conventional leaf-spring layout. Many of the earlier cars were converted to semi-elliptics by Ian Walker Racing but the original system can be made to work and parts are still available.

The steering shouldn't give much trouble, though the box does suffer from oil seal failure, which is a probable cause of undue stiffness.

Brakes are reliable, P16 calipers being fitted to the front discs, while 8-inch drums are used at the rear, as opposed to 7.5-inch ones on the Cortina GT.

Engine, gearbox and transmission

As the engine is shared with the Lotus Elan full details can be found on page 180.

Bucket seats, special instrument panel, wood rimmed steering wheel and remote control gear lever were all peculiar to the Cortina Lotus, in this instance a 1963 model. *Right*, under the bonnet

The gearbox hasn't a reputation for unreliability and it is probably worth mentioning that some early cars do have an aluminium bell housing. From mid-1964 a two-piece propeller shaft was introduced — a significant improvement on the earlier one-piece unit. The rear axle is durable.

Spares

Many of the components were common with the standard Mark I Cortina so you might still be able to locate such valuable items as replacement front wings via the usual trade sources. A number of specialist firms will undertake work on these twin-cam engines, just two being Ian Walker of Leighton Buzzard and Vegantune, of Spalding, Lincolnshire and you'll find full details on page 181. Membership of the rapidly-expanding Club Lotus is essential if you own one of these cars and a register of Cortina Lotuses is now well under way.

Cortina Lotus performance and consumption, *Autocar*, November 11, 1963.
Top speed: 106mph. Overall mpg: 20.8 for 1245 miles.

Ford Cortina Lotus production

1962		25*
1963		228
1964		567
1965		1118
1966		989
	TOTAL	2927

* estimate

William Lyons, educated at Arnold House School, Blackpool, whose father ran a piano business in the town, went into partnership at 21 years of age with William Walmsley to build Swallow sidecars for motorcycles. The year was 1922 and the overdraft of £1000 necessary was guaranteed by the partners' respective fathers. The business flourished and, in 1927, the Swallow Sidecar Company branched into the motor business. They produced their own 2-seater body for the Austin Seven chassis, and followed this in 1928 with a stylish saloon. That year the business moved from Blackpool to Coventry, at that time the very heart of the British motor industry. A former shell-filling factory in Foleshill was purchased and output of bodies spiralled. The stylish designs were applied to Standard, Swift, Fiat and Wolseley Hornet chassis.

The next stage was for Lyons to establish a marque of his own, but first he needed a name. He had a long wrangle with R. W. Maudslay and John Black, of the Standard Motor Company, as his car had been designed to carry their engines and running gear. Both parties eventually agreed on S.S., but whether it stood for Swallow Special or Standard Special was never resolved. The S.S.1 made its appearance at the 1931 Motor Show with a 16 or 20HP Standard engine. Even if the mechanics weren't likely to raise any eyebrows the car's looks most certainly were. The low lines and long bonnet proved a sensation, the *Daily Express* proclaiming it 'the car with the £1000 look for £350'. It was a theme that Lyons was to repeat again and again with devastating effectiveness. Here was a proprietor who personally styled his own products and his superb eye for line and proportion seldom let him down.

The S.S. had already gained a reputation for good looks, now it needed performance to complement them. In 1935 came the SS90, Lyons's first sports car and, like all his models at that time, it was powered by a side-valve Standard engine. The lines were impressive but the SS Jaguar 100, which appeared for 1936, offered looks *and* performance. Under the bonnet the bones of the Standard 2663cc engine, modified and fitted with a new overhead-valve cylinder head designed by Harry Weslake. It was also used in the $2\frac{1}{2}$-litre saloon, though the $1\frac{1}{2}$-litre version retained the side-valve layout. These models are also significant because they represent the first appearance of the Jaguar name, all cars thereafter being called SS Jaguars. The publicity department had prepared a list of bird, animal and fish names for Lyons as he felt his new overhead valve cars should have a model name. 'I immediately pounced on Jaguar because it had an exciting sound to me'. He also recalled that a friend had worked on a Siddeley Jaguar aero-engine during the First World War and he therefore asked Sir Frank Spriggs, then managing director of Armstrong Siddeley, whether he had any objection to his using the name. Lyons was given approval verbally, and in writing, though ironically, 23 years later, his permission was sought when it was decided to use the name for a new fighter strike aircraft!

William Heynes, who had joined SS Cars as chief engineer in 1935, further stretched the long-suffering 20HP Standard block to $3\frac{1}{2}$-litres for the 1938 season. At the same time the range of SS Jaguar saloons with all-steel bodywork was introduced. They were now available in $1\frac{1}{2}$-, $2\frac{1}{2}$- and $3\frac{1}{2}$-litre form, all with overhead-valve layouts. The 100 sports car also received the new $3\frac{1}{2}$-litre engine, giving it a top speed of just over the magical 100mph.

Therefore, at the outbreak of the Second World War, the successful birth and growth of SS meant that the name of William Lyons was already a force to be reckoned with. But he was, and today his company still is, first and foremost a producer of saloon cars.

During the war, the SS works at Holbrook Lane diverted their energies to the production of aircraft and aero-engine components, for which they built a sizeable new machine shop. Fuselages for the Stirling bomber and the Gloster Meteor, which was Britain's first operational jet, were two of the more notable products.

After the war Lyons decided to change the name of his company from S.S. Cars Ltd to Jaguar Cars Ltd. 'as initials SS had acquired a tarnished image, as it was a reminder of the German SS troops, a sector of the community which was not highly regarded'. Some months after the end of hostilities, Standard's Sir John Black informed Lyons that he could no longer supply him with 6-cylinder engines because their resources were to be concentrated on just one model, the Vanguard. But he agreed to sell, at a bargain price, the tooling for the work. Lyons quickly collected the machinery and paid for it, fearing that the quixotic Black would change his mind, which he promptly did! He proposed that they revert to their original arrangement and jointly form a separate company to produce cars, but Lyons would not agree. Later Lyons was approached with a view to his buying the moribund Triumph company, whose works adjoined his own, but he decided against it. This was a cue for Black to reiterate his original proposition but Lyons again declined. Black therefore told Lyons that he would buy Triumph and go into competition against him.

Lyons, meanwhile, was getting on with the business of producing his own cars. The first postwar Jaguars were essentially his 1938 range of saloons, though hypoid rear axles were fitted. In 1949 came a stop gap, the Mark V, with slightly altered coachwork, a new chassis fitted with a wishbone-and-torsion bar independent front suspension system designed by chief engineer Heynes before the war. The $1\frac{1}{2}$–litre option was dropped for this model, being available only in $2\frac{1}{2}$– and $3\frac{1}{2}$–litre form, the smaller capacity car being phased out in 1950 with the larger version remaining until the following year. Its replacement was the Mark VII, announced in October 1950; Mark VI was skipped because Bentley were already using the designation. This was a 5/6–seater saloon clearly designed for the American market. But above all it employed the fabled 3.4–litre 6–cylinder XK engine, which had been first shown to the public in the XK120 sports car that had caused such a stir at the 1948 Motor Show.

It was an instance of the well-established Lyons formula of sensational looks, astounding performance and excellent value for money. Lyons had always offered prices well below his competitors but, in 1948, there was an added incentive as cars costing more than £1000 were liable for double purchase tax at 66.6 per cent. The XK120, therefore, cost £988, plus £275 3s 11d purchase tax, which was exactly the same price as his Mark V $3\frac{1}{2}$–litre saloon. The XK120 was a significant milestone for the marque because it was the first model to be fitted with an engine of the company's own, rather than a Standard, design. William Heynes, along with Walter Hassan and Claude Baily had conceived the design under Lyons's watchful eye at fire-watching sessions during the Second World War, a fact revealed by Lyons in May 1949 when an XK120 covered the flying mile at 132.596mph on the Jabbeke–Aeltre autoroute in Belgium. It should not be forgotten that this engine was designed to propel a full-sized *saloon* car at

The XK engine, the heart of Jaguar's postwar success. Lyons always had an eye for visual appeal, so the stove enamelled exhaust manifolds make a striking contrast to the aluminium cam boxes

100mph in standard form, the XK120 sports car being created as a test bed for it.

The basis of the model was the Mark V chassis with 1ft 6ins removed from the frame. Lyons then created a magnificent two–seater body, clearly inspired by the Touring bodied BMW 328 which was built for the 1940 Mille Miglia. Initially the 120's bodywork was produced by traditional methods: by panelling an ash frame with aluminium sheet. Orders poured in for the new car catching the Foleshill factory by surprise and, though the first 240 sports cars were coachbuilt, the remainder incorporated panels produced by Pressed Steel, thus streamlining the production process.

In 1950 three XK120s, unofficially sponsored by the factory, were entered for the Le Mans 24–hour race. Although the Johnson/Hadley car dropped out after 21 hours while in third place and the other two entries finished in 12th and 15th positions, Lyons was convinced of the value of entering. As he recounted later: 'We decided that in a car more suitable for the race, the XK engine would win this greatest of all events.' The publicity incentive was enormous and the 'more suitable' car was William Heynes's tubular chassis C type, with a team of three entering for the 1951 event. Although two of the cars dropped out, Peters Whitehead and Walker went on to win at a record speed of 93.49mph. It was the first British win at Le Mans since Johnnie

Hindmarsh and Luis Fontes took the chequered flag for Lagonda in 1935. The C type's victory set the pace for Jaguar successes at the Sarthe circuit, taking first place on no less than five occasions — 1951, 1953, 1955, 1956 and 1957. Perhaps the greatest year was 1957, ironically after Jaguar had officially withdrawn from racing, when Edinburgh-based Ecurie Ecosse-entered cars took the first two places. Privately entered Jaguars finished third and fourth. The C type held sway until 1953, and in 1954 it was replaced by the Malcolm Sayer-styled D type, which favoured a monocoque hull. A road-going version of the D, the XKSS, was offered briefly for public sale in 1957.

By contrast, the XK range of sports cars was produced in much larger numbers. The 120 ceased production in 1954 and was replaced by the XK140 which retained the same basic body style, though benefiting from rack-and-pinion steering (as proven on the C type) and more substantial bumpers. This had a 3–year production run for in 1957 came the XK150 with its updated body lines and all-round disc brakes. All the XK range were fitted with a 3.4–litre engine but when the 3.8–litre unit became available for 1960 it was offered as an alternative fitment in the 150 model. Such was the growth of the company that Jaguar were rapidly outgrowing their Foleshill premises and in 1951 production was transferred to the former Daimler Shadow Factory No. 2 at Browns Lane, Allesley on the outskirts of Coventry. Lyons was able to buy, rather than rent the works as was the normal case with such establishments. It did mean that the company had to undertake the manufacture of the Rolls–Royce Meteor tank engine, though this proved to be a short-lived arrangement.

Lyons's main source of revenue was his saloon cars and a new model appeared for the 1956 season. This was the unitary construction 2.4 powered by a short-stroke version of the XK engine, though it had coil-spring-and-wishbone independent front suspension in place of the established torsion-bar system. At the rear cantilever springs and radius arms were employed. Sales in the USA were so good (exports never dropped below 50 per cent of production during the period under review) that the company offered the 3.4–litre engine in the 1957 car. The Mark II version of the design, available in 2.4–, 3.4– and 3.8–litre forms, appeared for 1960. Its larger brother, the Mark VII, remained in production until 1956 when it was replaced by the

The Autocar road tested this car in 1951, a year in which 96 per cent of Jaguar's output was exported. The token rear bumpers do little to detract from the magnificent lines

more powerful Mark VIII for 1957. The 1959 Mark IX featured a 3.8–litre engine and disc brakes, and power-assisted steering was a standard fitment. It was phased out in the autumn of 1961.

Jaguar were again outgrowing their premises and Sir William (as he became in 1956) did not want to extend outside Coventry, though government policy favoured factory development in high-unemployment areas. In 1960 he heard that Daimler, with a fine factory at Radford just 2 miles away was up for sale and he purchased the company from BSA for £3.5 million. Commercial vehicle manufacturers Guy Motors of Wolverhampton were bought the following year while Coventry Climax, who produced the fork-lift trucks and fire pumps, entered the Jaguar fold in 1963. Finally, Henry Meadows of Wolverhampton, the well-established manufacturers of gearboxes was purchased in 1964.

Meanwhile, on the manufacturing front, 1961 was extremely significant for Jaguar. In March the company's best-selling sports car, the E type, was announced. The sleek, stylish and utterly distinctive body was the work of Malcolm Sayer and in concept sprang from the Le Mans winning D type. It was, therefore, a true monocoque, but with a new independent rear suspension system, while at the front it retained the D's torsion bar-and-wishbone layout. Then, in October, a replacement for the ageing Mark IX was announced. The new Mark X was, like the E type, powered by a 3.8–litre version of the XK engine and inherited its independent rear suspension. This saloon, with its fashionable four-headlamp layout, was also of monocoque construction.

In 1963 a model combining the Mark X's independent rear suspension and tail and the Mark II's body style was announced. The S type, identifiable by its heavier radiator grille surround and hooded spot and headlamp shrouds, was available in 3.4– or 3.8–litre forms and remained in production until 1968. A new 4.2–litre version of the faithful XK engine was fitted to the E type and Mark X for 1965 with an all synchromesh gearbox appearing at the same time. The proprietary robust, but notchy, Moss gearbox from 1966 (on the Mark II) ceased to be a Jaguar fitment. The 420 was announced for the 1967 season. It was, in effect, the S type saloon with Mark X front, that model being designated the 420G and the following year the Mark II 2.4–litre was rationalized as the 240 model and the 3.4–litre the 340.

Jaguar Cars' independent existence came to an end in 1966 when the company merged with the British Motor Corporation to form British Motor (Holdings). It was a move that Sir William believed was both 'necessary and desirable'.

Jaguar production 1945–1967

$1\frac{1}{2}$ litre saloon		5750
$2\frac{1}{2}$ litre saloon		1850
$3\frac{1}{2}$ litre saloon (including drophead coupé)		4400
Mark V $2\frac{1}{2}$ litre saloon		1661
Mark V $2\frac{1}{2}$ litre drophead coupé		29
Mark V $3\frac{1}{2}$ litre saloon		7831
Mark V $3\frac{1}{2}$ litre drophead coupé		972
	TOTAL	10,493
Mark VII (all types)		31,000
C type		54
D type		62
XKSS		16
Mark VIII saloon		6300
Mark IX saloon		10,000
Mark X saloon (3.8)		13,000
Mark X saloon (4.2) including 420G		10,550
	TOTAL	23,500
420 saloon		9800
S saloon (3.4)		11,050
S saloon (3.8)		15,150
	TOTAL	26,200

XK120 1949–1954
XK140 1955–1957
XK150 1957–1961

Specifications and developmental history

The XK120 was announced as an open 2–seater at the 1948 Motor Show. A box-section chassis with torsion bar-and-wishbone independent front suspension was employed. Lockheed hydraulic drum brakes and Burman re-circulating ball steering were fitted. The engine was a 3442cc (83 × 106mm), 6–cylinder twin overhead-camshaft unit developing 160bhp. Seats were upholstered in leather and

Overleaf, **a 1953 XK 120 drophead coupé. Announced in 1953, it was the last variation of the range and at £1140 was only initially available in America**

the dashboard was finished in the same material. The car cost £1263. In March 1951, a fixed-head coupé version, of similar specifications to the open two-seater, but with walnut facia and door cappings appeared. From August 1951 a 180bhp Special Equipment engine was available for all XKs. The package included $\frac{3}{8}$–inch lift camshafts (in place of the standard $\frac{5}{16}$–inch) and a dual exhaust system. A drophead coupé version, sharing the same interior fitments as the fixed head model, appeared in April 1953.

The model was replaced in October 1954 by the XK140, available in open 2–seater, fixed and drophead coupé forms. All but the open car were fitted with two occasional rear seats for children so the two 6–volt batteries were moved to the front wings. The engine, which was moved forward 3 inches in the chassis, was the Special Equipment version of the XK120 unit, while the C type head used for racing was also available. This had been a fitment on the XK120 from the spring of 1953 and differences included increased diameter inlet and exhaust porting and larger exhaust valves. The head, which can be identified by the letter C cast in the plug wells, contributed to an ouput of 210bhp. Steering was a departure from previous practice

XK 140 coupé. Centre chrome boot strip and more substantial bumpers distinguish this model from the XK 120. The wire wheels were an optional extra

Driving compartment of a 1951 XK 120. The dashboard is leather covered. Note fly off hand brake

An XK 150 drophead coupé. The smoother body lines and wider radiator grille instantly identify the model

being an Alford and Alder rack-and-pinion unit. The car was externally distinguishable from the XK120 by Mark VII-type bumpers and the coarser and fewer slatted radiator grille, while the bonnet was decorated by a chromium strip along its centre. Wing-mounted flashing indicators were fitted, while the rear ones were incorporated into the back lights. A chrome strip on the boot lid matched the one on the bonnet.

The XK150 was announced in fixed and drophead coupé form in May 1957. The model was noticeably different from earlier versions of the XK theme with a higher scuttle and door line, while the radiator grille reverted to the thinner slats of the XK120, but was altogether wider. Larger wrap-around bumpers were fitted and a curved windscreen replaced the divided one on the XK120 and 140. Servo-assisted Dunlop disc brakes were fitted all round. The engine was the standard 140 unit, though a 210bhp Special Equipment version was also available. This B type head retained the large exhaust valves of the C type but utilized the smaller inlet porting of the standard head. Wood door cappings and facia were absent from all versions, being replaced by leather. The XK150 open 2-seater appeared in March 1958 in standard and S form. The latter variant used a new 'straight port' cylinder head, triple 2-inch SU carburettors, instead of the twin 1.75-inch ones and 9:1 compression ratio in place of the 8:1 option which was more usual. In February 1959 this more powerful engine was extended to the fixed and drophead coupé models. In October 1959 a 3.8-litre (87 × 106mm) version of the XK engine was announced with a choice of heads, the twin carburettored B-type (220bhp) or triple carburettored S-type (265bhp). Manual gearbox and overdrive was a standard fitment on all S models, otherwise automatic transmission was available as an optional extra. In view of the complexity of the engines available, the engine numbers were prefixed as follows: V=3.4, VS=3.4S, VA=3.8, VAS=3.8S. All models were discontinued in October 1961.

Check points

Body and chassis

If you're lucky enough to find one of the first 240 XK120s then the following details won't apply, as they used a traditional ash-framed body clad in aluminium. Any shortcomings will probably be associated with the ash frame rotting but the cars are highly 'collectable' so don't let this deter you! Incidentally, the last left-hand-drive aluminium XK120 was chassis number 670184 and the final right-hand-drive car was 66058, dispatched in April and May 1950 respectively. (66 indicates rhd and 67, lhd.) From thereon steel took over. So, starting your investigations at the front of the car, feel for rust at the bottom of the wings beneath the air intakes, as corrosion can get a grip. Other vulnerable points are around the head and side-lamps. Check along the bottom of the wings and also around the side ventilator and its flap. On the XK140 fixed and drophead coupés and XK150s the batteries were carried in the front wings so look for corrosion around the battery box, though the inner wings forward of them are usually sound.

The doors can give trouble and may have suffered from past 'repairs'. Hinges are a problem and the difficulty is that the wings have to be butchered to get to their boxes. To check their condition, slightly open each door in turn and move it up and down. You'll soon be able to detect any excessive wear. Lifting the door in order to close it or its reluctance to remain closed are both indicative of more expense. The doors themselves have a tendency to rust from the bottom upwards. Watch for evidence of plating: if there are no drain holes in the bottom of the doors this is a sure indication of such a repair. The lower portion of the door shut face also tends to rust badly.

The sills are particularly important in the XK's construction. They are built up in two halves and spot-welded together and this is often the first place to give trouble, the joint running the whole length of the sill. Check the points at which they join the lower edge of the door shut faces and inside the rear wheel arches. The ends should be protected by steel splash plates, but if these have rotted away their protective role is nullified and the sills start rusting from the inside. Examine the joint between the wings and the body sides and 'topsides look for rusting either side of the wing beading. On the fixed-head coupé cars a danger point is just beneath the rear-side windows. Inevitably the boot is another rust trap. Warping of the lid's wooden frame is the usual reason for a bad fit and check the spare wheel tray, particularly along the seams.

Chassis problems are virtually unknown on the XK series and the only area that might give trouble is where the frame sweeps directly over the rear axle.

Suspension, steering and brakes

The front suspension is notably uncomplicated, but check the upper and lower ball joints for wear.

A leaking radiator can, over the years, cause rusting in the front suspension mountings, but this is unusual. The cart-sprung rear suffers occasionally from broken leaves.

Steering was re-circulating ball on the XK120, though rack and pinion was fitted from the 140 onwards. The latter system is obviously more desirable and neither have many snags, though the rack shouldn't slacken and the rubber mountings must be sound.

Similarly, the XK150's Dunlop disc brakes were a great improvement on the drums previously employed, though spares for the round-padded versions are becoming scarce. They're also more complicated to fit than the later square-section ones as the calipers must be dismantled. Lockheed servo assistance is another advantage. These discs can be fitted to the earlier cars to replace the drums which were susceptible to wear and they do nothing to mar the car's appearance. However, if originality is paramount, then you'll probably leave things as they are.

Engine, gearbox and transmission

All XKs were produced with 3.4–litre engines, though some 150s were given 3.8–litre units, while the 'straight port' head with triple 2-inch carburettors instead of the two 1.25-inch SUs was a desirable extra. Because engines are so readily obtainable always check that the engine number corresponds with the chassis plate and log book.

The engine is very reliable and its longevity is legendary — 100,000 miles is quite possible without major overhaul. The best way to ascertain its condition is to drive around in top gear at about 40–50mph for a while and then accelerate sharply. If the power unit is in good condition there should just be a light haze from the exhaust but clouds of blue smoke indicate bore wear. Ideally oil pressure should not fall below 35–40psi (hot) at 3000rpm and listen for bearing noise. Some tappet adjustment may be necessary, but apart from this the top end should be fairly quiet. There are two timing chains with wear in the lower one being the more serious of the two.

You shouldn't have any problems with the clutch, though it is usually well worn after 30,000 miles. The Moss gearbox is a sturdy unit but beware noisy first and reverse gears as spares are no longer available. Borg–Warner automatic transmission was a fitment on the XK150 but parts are also becoming tricky for this and the same goes for those cars fitted with Laycock overdrive.

You shouldn't suffer from rear axle problems with these cars.

Spares

XK owners really are well catered for when it comes to practically all types of parts, be they body, trim or mechanical. Even difficult exterior trim items are reproduced, both in Britain and America and in many instances they are indistinguishable from the original. Just two companies that supply spares for them are Classic Cars of Coventry and Oldham and Crowther of Peterborough. Classic Autos of King's Langley, Herts also specializes in body parts. A London source of Jaguar parts are G. H. Nolan while Suffolk and Turley of Nuneaton specialize in re-trimming these cars. The large and well-run Jaguar Drivers' Club has numerous sections to cater for the individual models: XK, E type, Mark II, etc. and membership is therefore a 'must'.

XK spares and services

Classic Autos, 10 High Street, King's Langley, Hertfordshire (tel: 09277-62994).
Classic Cars of Coventry, Jaguar Corner, Southfield Road, Hinckley, Leicestershire (tel: 0455-613948).
G. H. Nolan Ltd., 1 St George's Way, London SE15 (tel: 01-701 2785/2669).
Oldham and Crowther, 27–31 Westwood Industrial Estate, Peterborough, PE3 7PH (tel: 0733 265021/265046/262577)
Suffolk and Turley, Unit 7, Attleborough Fields Industrial Estate, Garrett Street, Nuneaton, Warwickshire (tel: Nuneaton 381429).

Jaguar Drivers' Club, Norfolk Hotel, Harrington Road, London SW7 (tel: 01-584 9494/5).

A 1958 3.4 litre temporarily ignoring the parking instructions

Performance and consumption

Jaguar XK120, *The Motor*, November 16, 1949. Top speed: 124.6 mph. Overall mph: 19.8 for 1020 miles.

Jaguar XK production

XK120		
Open 2-seater		7612
Fixed head coupé		2678
Drophead coupé		1765
	TOTAL	12,055
XK140		
Open 2–seater		3354
Fixed head coupé		2808
Drophead coupé		2889
	TOTAL	9051
XK150		
Fixed head coupé (3.4)		3445
Drophead coupé (3.4)		1903
Open 2–seater (3.4)		1297
S open 2–seater (3.4)		888
S fixed head coupé (3.4)		199
S drophead coupé (3.4)		104
Open 2–seater (3.8)		42
Fixed head coupé (3.8)		656
Drophead coupé (3.8)		586
S open 2–seater (3.8)		36
S fixed head coupé (3.8)		150
S drophead coupé (3.8)		89
	TOTAL	9395

Total XK production: 30,501

attained a top speed of 120mph, in overdrive, in this car

2.4 litre 1956–1959
Mark II 1960–1969

Specifications and developmental history

The 2.4–litre Jaguar appeared at the 1955 Motor Show. It was a monocoque 4–door saloon with coil-and-wishbone independent front suspension, full cantilever springs at the rear, the axle being located by radius arms and a Panhard rod. Servo-assisted, Lockheed hydraulic drum brakes were fitted and Burman re-circulating ball steering was used. The engine was a new version of the XK 6–cylinder, twin overhead-camshaft series of 2483cc (83 × 76mm), developing 112bhp. Seats were upholstered in leather and door cappings and facia were finished in walnut. The basic model cost £1269. Early in 1957 the 210bhp, 3.4–litre engine was fitted, and the car was identifiable by a wider radiator grille, absence of spot lamps and cutaway rear spats. Automatic transmission was an optional extra. In September the new grille was added to the 2.4–litre version and Dunlop disc brakes were optional on both models.

A Mark II version appeared in October 1959. The new body was instantly recognizable by a modified radiator grille with a more pronounced central rib and separate partly-recessed fog lamps. Side-lamps were mounted on top of the wings, flashing indicators occupying their position. Window space was increased, the facia was re-styled and the seats fitted with deeper squabs. Mechanically, the rear axle width was increased by 3.25 inches and servo-assisted Dunlop disc brakes were standard. It was offered in 2.4–, 3.4– and also 3.8–litre (220bhp) form. From September 1960 power steering was available on the two larger capacity cars. Reclining seats were optional. A new all-synchromesh gearbox was available in 1966. The range was revised in September 1966 — the fog lamps were dropped, while tufted carpets replaced the earlier pile ones. Ambla upholstery replaced leather, though leather was still available at extra cost. The automatic gearbox option was dropped. For the 1968 season the model's designation was changed to 240 (2.4 litre) and 340 (3.4 litre) and new slimmer bumpers were introduced. And the automatic gearbox option was reinstated. The smaller capacity model remained available until April 1969, the larger one having been dropped in September 1968.

The Mark II version, which was available in 2.4, 3.4 and 3.8 litre variants. The narrower window pillars, inbuilt spotlamps and more substantial bumpers are identifying features of this 1960 car

Check points

Bodywork

Start your investigations at the back of the car and if this area doesn't pass muster then pass on to another example. First, check the state of the boot lid, which is double-skinned. If this is satisfactory, open it, remove the matting on the floor and look in the boot extremities where the rear dampers are mounted. If there are signs of corrosion here, be particularly suspicious. If possible, ascertain the condition of the metal towards the tail, and level with the floor; this, however, is easier said than done as it is normally obscured by cardboard trim. If corrosion is present it can spread to the wheel arch seams, which is a further disadvantage. Next, remove the spare wheel lid and lift out the wheel to examine the well, though its condition isn't crucial. However, your next area for investigation is. The state of the spring hanger boxes containing the rear cantilever springs is vital. If these boxes are badly rusted, the vehicle is far from safe, but a repair is possible, though probably only worthwhile if everything else is in order. Check also the condition of the Panhard rod mountings as these can often be troublesome. Finally, check the corresponding interior and, if allowed, remove the rear seat squab to facilitate careful inspection of the edge of the seat pan. Raise the carpet where the pan meets the floor and examine for evidence of rusting. If these two final checks show that rust is established, it is advisable to look for another example. Complete the inspection with a look at the partially double-skinned rear valance under the bumper.

Next—the doors. These can rust and this is often caused by blocked drain holes. Remember to check about half-way down the door as water collects in an internal pad, which cushions the window winding mechanism.

At the front examine the valance, particularly where it runs beneath the radiator grille. Also an internal valance, which carries the bumper, can rust particularly badly from mud thrown up by the road wheels. Next ensure that the vacuum tank and shield on the wheel arches are intact on the offside and check the condition of the cross-member, which is forward of the engine. Mud also gets trapped on the lips on the edges of the front wings and where the wings meet the closing plate behind the wheels. Damage here admits water and dirt into the sills which soon causes internal rusting. Corrosion of the wing at the bottom of the bulkhead is probably due either to damage or perishing of the rubber strip between the wings and the closing plate.

The sidelight housings should be examined, as they tend to disintegrate and corrosion can also be found in and around the flasher bowls. You can complete your inspection by looking at the outer sills. Their condition will probably be apparent and you have already investigated the inner ones.

Suspension, steering and brakes

The front suspension units are carried on a subframe and, if the front of the car looks unduly low, this is attributable either to weakened coil springs or deterioration of the rubber subframe mountings. Fortunately, both can be rectified without too much difficulty. Otherwise, providing it has been properly maintained, the front suspension shouldn't present any major problems. If not then you'll probably find some wear in the upper and lower ball joints. Shimming is employed in the bottom joint, which is an advantage, but if the car is suffering from heavy steering, it might be caused by a build up of rust.

Steering gear, if well maintained, is satisfactory, though oil from the engine can perish the rubber bonding on the track rod joints. Power assistance was an optional extra on the model and a noisy, leaky box invariably indicates trouble.

Brakes are mostly trouble-free and any problems are usually associated with the disc variety. A rusty disc means that a wheel cylinder isn't working, while a seized piston can cause a disc to break up if the car is run for any length of time. The handbrake isn't spectacular and its poor action is usually caused by seized fulcrum bolts which are situated near the rear brake calipers. A heavy brake pedal is probably caused by trouble with the servo, which may have a rusty reservoir.

Engine, gearbox and transmission

Engine checks are similar to those recorded for the XK range. It is probably worth noting that, while the 3.8– and 3.4–litre engines used SU carburettors, twin Solexes were employed on the 2.4.

There is a higher incidence of problems on the earlier gearboxes which were not all-synchromesh. These were rather noisy in first and reverse gears anyway, but the general absence of spares of first and reverse layshaft does not help the overall situation.

Rear axle problems are virtually unknown but you should note that each engine employed its own rear axle ratio and, if a power unit has been changed, then ensure that the axle was altered at the same time.

Spares

The availability of mechanical parts is the same as for the XK range (see page 154), but, until recently, body panels have been a problem. Replicas are now available from a number of companies including PSW Panels of Coventry, while Peter Dee of Birmingham concentrates specifically on the Mark I and II range.

Mark I/II spares and services

Peter Dee Ltd., 64 Mount Street (off Louise Lorne Road), Moseley, Birmingham 13 (tel: 021-449 6082).

PSW Panels, 76a Albany Road, Earlsdon, Coventry (tel: 0203-74030).

Interior of a 1958 3.4. Leather and wood is used to great effect. The steering wheel has telescopic adjustment

Above, an early Mark II engine bay. Up until 1960 an oil bath air filter was used. From thereafter a round paper element was employed. *Right*, superb lines. A 3.8 litre E type, one of the greatest postwar Jaguars

Performance and consumption (Mark I, 3.4–litre with overdrive), *The Autocar*, June, 13 1958.
Top speed: 120mph. Overall mpg: 16.5 for 2121 miles.

Jaguar Mark I and Mark II production

2.4–litre		20,000
3.4–litre		17,350
	TOTAL	37,350
Mark II 2.4–litre		30,600
Mark II 3.4–litre		31,450
Mark II 3.8–litre		30,510
	TOTAL	92,560

E type 1961–1971

Specifications and developmental history

The E type Jaguar was introduced in fixed-head coupé form at the Geneva Motor Show in March 1961. Also available as an open 2–seater, the E type was a monocoque with a front triangulated subframe. Suspension was all-independent with a wishbone-and-torsion bar layout at the front, and twin coil springs used in conjunction with a lower tubular link at the rear. Brakes were servo-assisted Dunlop discs all round. Rack-and-pinion steering was employed. The engine was the 3.8–litre unit as fitted to the XK150S, developing 265bhp. Unpleated leather bucket seats were fitted. The coupé was priced at £1550, while the open two–seater sold for £1480. Changes to all but the earliest cars included water deflector shields to the front axle carrier, interior bonnet-release catches and, in October 1961, a self-adjusting handbrake was introduced. The following year, in July, a one-piece, forged steering column replaced the tubular shaft assembly. In June 1963 thicker 0.5-inch-diameter rear brake discs and new calipers were introduced. A Laycock diaphragm clutch appeared in March 1964.

In October 1964 a 4.2–litre (92 × 106mm) engine, developing 265bhp, replaced the 3.8–litre unit. Other differences were: twin SU (series 1) petrol

pumps in place of a Lucas pump immersed in the tank; a Lucas alternator was introduced and electrics accordingly changed to negative earth; a pre-engaged starter was employed; a Lockheed vacuum servo replaced the Kelsey Hayes unit originally fitted; seats were re-designed, being plusher and pleated; and an all-black instrument panel replaced the original aluminised one. A major improvement was the fitting of a new Jaguar-designed all-synchromesh gearbox. The rear axle ratio was changed from 3.31:1 to 3.07:1. March 1966 saw the appearance of the 2+2 fixed-head coupé with 8ft 9in wheelbase instead of 8ft, and longer doors. A wide-ratio gearbox was specified, while automatic transmission was an optional extra. Late in 1967 came what is unofficially

A 4.2–litre fixed head coupé. Shorter silencers were used compared with those fitted on the earlier 3.8 model

known as the Series 1½ distinguishable by the absence of Triplex headlamp covers and rocker switches on the instrument panel, and key starting was a new feature. Under the bonnet was a new cross-flow radiator and twin electric fans.

In October 1968 the Series 2 E type was announced, the changes being made in deference to the American safety regulations. Larger, raised bumpers were fitted, while the open headlamps were advanced 2 inches. Repositioned larger sidelamps and flasher units were introduced and the bonnet aperture was enlarged. Twin reversing lights were standardized and power steering was an optional extra. During 1969, new camshafts were introduced for quieter running, and a steering-column lock, gas filled bonnet stay and arm rests on the doors were other new features in this period. The following year the wire wheels were fitted with non-eared safety hub caps, while disc wheels were optional. A Series 3 6–cylinder E type was offered when the 5.3–litre V12 version was introduced in 1971, but only two or three fixed-head examples of the smaller-engined car were produced. Production ceased in 1971, while the V12, which employed the long wheelbase body, remained in production until 1975.

Check points

Bodywork

Starting at the front of this handsome vehicle, first examine that elegant but vulnerable bonnet. Begin by gently tapping it; a heavy resonance probably indicates filler and, by placing your hand inside the bonnet, it should be possible to detect the filler. Should you disturb rust in the lower nose section this probably indicates advanced deterioration of the panel underneath. Another rust spot is on the top of the wings adjoining the chrome strip, which marks the joint between them and the bonnet centre. You can assess the state of this joint by opening the bonnet and looking underneath; often the flange joint is rusted badly. Stand well back from the car and see how well the bonnet fits. Poor alignment may betray a past accident though shims are provided on the hinges and adjustable flanges at the bonnet catch end.

Move on to the main body itself. The original sills, if they are still present, were spot-welded into position, but replacements will probably have been continuously welded or brazed. The condition of these sills is vital to the general well-being of the car, particularly on the Roadster which doesn't have the support of a metal roof, and undue flexing of the body can result if the sills are badly rusted or holed. Examine the rear wing edges and also underneath in the triangular area formed by the bumper just behind the rear wheel. The Roadster model doesn't have an inner wing so look for evidence of rusting on the apex of the wing proper. However, on the Fixed Head E type, probe the inside of the rear wings and around their edges for any signs of rusting.

Next, open the boot and remove the spare wheel from its well, paying particular attention to the condition of the point at which the inner wing joins the floor. Also inspect underneath the car at the three stiffeners running beneath the floor. Check around the rear number plate and also the rear panel section above the exhaust pipes. Whilst inspecting the underside check the points at which the rear radius suspension arms are attached to the floor. Any rusting here is crucial and, in severe cases, can be thoroughly dangerous. Check also the state of the hollow transverse beam that runs beneath the seats.

The principal problem with the doors is corrosion, particularly where the outer skin joins the inner section. In order to determine if the rust has also affected this area, grasp the bottom edge with your thumb on the exterior panel. Then try to flex the metal. An ominous but muted cracking noise will confirm advanced corrosion.

A particularly important component to check is the front subframe. Two major problems should be considered when conducting the inspection: the inevitable rust and signs of accident damage. Examine carefully all the joints and look for denting on the bulkhead (the sign of a bad crash) at the points where the subframe is bolted to it. The E type chassis number on the front cross member can be found above the damper mounting. If this is absent a new subframe has been fitted at some stage.

Suspension, steering and brakes

The E type independent front suspension is remarkably trouble-free and wear is usually confined to the top and bottom suspension ball joints. Unfortunately, the same cannot be said for the rear! and any trouble here is usually audible to an alert test-driver, who can detect any ominous clunks or clicks by the accelerator pedal backing on and off. These could be symptomatic of a variety

Cockpit of a 1961 coupé. The unpleated seats gave way to pleated ones when the 4.2 was introduced in 1964

of shortcomings including worn universal half-shaft joints, slack in the rear-wheel stub axle splines or wearing of the bottom pivots on the stub axles. Some brisk cornering should reveal the state of the Metalastik differential bushes and if the car 'settles' or 'wallows' their condition could pose problems. A final test should be made while cruising along in top gear: any audible whine will be coming from the rear axle.

With the car stationary, loosen off the wheel nuts of both rear wheels, put the car in gear and apply the handbrake. Now jack the car up and, with your hands in the nine and three o'clock positions, move the rim within a limited arc. Discernible movement will confirm that the wheel splines are worn and this will have to be rectified. Be suspicious of the front wheels too! Visual inspection of the splines can be made by removing the wheels. The splines should be flat on top but if they are pointed then this indicates excessive wear.

A leaking differential housing is fairly common on E types and other independent rear suspension Jaguars. This is caused mainly by overheating of the drive, the heat being generated by the inboard disc brakes. Also the drive shafts serve as the top suspension links, which puts an additional strain on the oil seals. Lubricant therefore finds its way on to the inside faces of the discs which inevitably detracts from the brakes' effectiveness. A clean breather does prevent pressure building up in the casing, however, but the only solution is to dismantle the final drive assembly and fit new seals. The differential unit itself is mounted on to the body via large Metalastik mountings. To ensure that these important components haven't separated, jack the car up and place it on axle stands positioned on the two longitudinal body members, but with the jack positioned under the differential unit. Lowering the jack slightly will reveal if the metal and rubber parts have separated.

Other potential points of wear are the bottom pivot bearings on the rear stub axle carriers, the inner needle roller bearings which are found on the inboard end of the wishbones, while the anti-roll bar mountings and bushes can also be affected.

The E type's steering is by rack and pinion and consequently should suffer from very little play. But if it does, first ensure that it isn't caused by wear in the two universal joints in the steering column. If not, then the rack could be moving on its rubber mounting bushes. The track rod ends should also be checked for wear.

Apart from the aforementioned problems with the inboard rear discs, the brakes shouldn't prove troublesome. Braking certainly improved over the model's life though it is probably worth mentioning that the Kelsey Hayes booster on pre Series 1 cars allows the car to run out of breath somewhat in the last moments of pulling up.

Engine, gearbox and transmission

You want to be certain that the E type under inspection is fitted with its correct engine. It is not unknown for an E type to contain the power unit from a Mark X saloon, since there was very little difference between the two specifications. The real problem is when a 3.4–litre motor with pre-straight porthead has been substituted. The engine number is located on the right-hand side of the block above the oil filter and also on the front of the cylinder head and these should correspond. These should be compared with those listed on the data plate located on the sill in the engine compartment.

Other checks are the same as those detailed for other XK engines (see page 154).

Clutches have a life of around 30,000 miles but where they have worn replacement involves removal of the engine. All too often though the job was bodged by removing the glass-fibre gearbox cover and cutting the metal underneath. This meant that the gearbox could be removed backwards, thus revealing the clutch and making engine removal unnecessary. Be alert for signs of such 'repairs'. The E type gearbox is reasonably trouble-free but some qualifications are necessary. The 3.8–litre version used the XK-derived Moss gearbox which has no synchromesh on first gear. It may whine somewhat in first, second and third gears, though top should be quiet enough. By contrast the all-synchromesh gearbox on the 4.2–litre should only whine slightly in first and second, points worth remembering.

Spares

Increasing numbers of replacement E type body parts are becoming available but the nature of the car's construction is such that if major surgery is contemplated then the shell will have to be mounted on a jig to prevent distortion. Phoenix Engineering of Goole, Yorkshire carry a large stock of E type parts as do the aforementioned G. H. Nolan and Oldham and Crowther (see page 154). Other E type specialists are Kenilworth Motor Racing Services of Hampton-in-Arden Warwickshire, Jim Tester of Daventry and Cliffe Motors of Leeds.

Under the bonnet of a 1961 3.8. Accessibility was good (not always a Jaguar strong point), even the hydraulic fluid was carried in transparent containers. Note the bellows of the Kelsey Hayes vacuum servo. A remotely mounted Lockheed unit replaced it on the 4.2 cars

E type spares and services

Cliffe Motors, Low Street, Sherburn-in-Elmet, near Leeds.
Kenilworth Motor Racing Services, Kenilworth Road, Hampton-in-Arden, Warwickshire.
Jim Tester, 3 School Road, Woodford Halse, Byfield, Daventry.
Phoenix Engineering, Rawcliffe Bridge, Goole, Yorkshire (tel: Goole 83339).

E type performance and consumption. (4.2 litre) *Motor*, October 31, 1964.
Top speed 150mph. Overall mpg: 18.5 for 3020 miles.

E type production

Open 2–seater (3.8–litre)		7827
Fixed-head coupé (3.8–litre)		7669
	TOTAL	15,496
Competition lightweight		12
Open 2–seater (4.2–litre)		9548
Fixed-head coupé (4.2–litre)		7770
2+2 (4.2–litre)		5598
	TOTAL	22,916
Series 2 open 2–seater		8627
Series 2 fixed-head coupé		4855
Series 2 2+2		5326
	TOTAL	18,808
Total 6–cylinder production (excluding lightweights)		57,220
Series 3 V12 open 2–seater		7990
Series 3 V12 2+2 coupé		7297
	TOTAL	15,287

Jensen 1936-1976

Alan and Richard Jensen (born 1906 and 1909 respectively) were sons of a provisions importer of Danish descent and were born in Moseley, Birmingham. Both entered the motor trade after education at King's School, Birmingham, Alan joining Serck Radiators and Richard serving an apprenticeship with Wolseley Motors.

Their first attempt at coachbuilding (this was to prove their forté) was on an Austin Seven Chummy chassis after being motivated by an appalling fearsome home-brewed special encountered at their local tennis club. Their next venture was, at the request of the Standard Motor Company, a further 2–seater body constructed on the Nine chassis. Via an introduction from Montague Tombs, Midland Editor of *The Autocar*, Alan joined Avon bodies to design the production version of the Standard Avon. In the meantime, Richard had left Wolseley and continued his apprenticeship at Joseph Lucas Ltd.

But by this time both brothers were determined to produce their own car and fortunately they were introduced to J. A. Patrick, who owned Edgbaston Garages Ltd. The industrious pair were subsequently invited to join the business. Inevitably, they established a coachbuilding section and soon joined the board of directors. The servicing side of the business was revitalized, but they overstepped the mark when they re-named the concern Jensen–Patrick Motors. This impertinence cost them their jobs, though they soon found new employment. A contact with their next-door-neighbour resulted in a meeting with George Mason, whose father ran a provisions business in Birmingham, and who also had an interest in a ramshackle coachbuilding firm called W. S. Smith and Sons, situated in High Street, Carters Green, West Bromwich. In 1931 Alan and Richard therefore joined the business and became joint managing directors.

The Jensen brothers succeeded in revitalizing the concern, which at the time was committed to the production of commercial vehicle bodies and inevitably Richard soon built up a car-body shop in a corner of the works. In 1934 the company's name was changed to Jensen Motors Ltd., and at this time their special bodywork was fitted to Morris, Singer, Standard and Wolseley Hornet chassis. They were also responsible for the bodywork of Ron Horton's MG Midget and Magnette single-seaters. Higher up the price range was a stylish Hudson fixed-head coupé for Carl Skinner of SU carburettor fame. The company achieved a big breakthrough when they built an open four–seater touring body on a Ford V8 chassis for film star Clark Gable and it was displayed at the 1934 Ford Motor Show, held at the Albert Hall. This attracted orders for about 20 replicas, though Jensens weren't able to obtain the chassis until Henry's son, Edsel, had approved the project after driving Gable's car.

Work on the prototype Jensen began in 1934. It had a specially built Rubery Owen chassis, Ford V8 engine and a Columbia 2–speed rear axle that gave six forward speeds. Fitted with a 4–seater touring body, this model was known as the White Lady. However, the car was displaced by a more traditional model, the S type 4–door saloon, which shared the same mechanics as the White Lady. Sales were encouraging, so the original touring design went into production for 1936 with minor modifications. Although they experimented with an 8–cylinder Steyr-engined car it never went into production; a 4.2–litre model with a straight eight Nash engine was more successful, and this model, which featured a longer chassis, was designated the H series. An HC type with coil-sprung rear suspension appeared in 1939, a few being fitted with the Lincoln Zephyr V12 engines for American customers.

Car manufacturing came to an abrupt halt at the outbreak of the Second World War and, during hostilities, specialist military vehicles such as ambulances and fire tenders were produced along with amphibious versions of the Sherman tank.

Postwar projects included bodies for Lea–Francis and the Invicta Black Prince, while the company's own model, to be made in saloon and drophead coupé versions, was announced in 1946. This was the PW (for postwar) with a chassis developed from the pre-war H type and with coil-and-wishbone independent front suspension. However, there were problems with the Henry Meadows 3.8–litre engine, so the first four PWs were converted to take pre-war Nash power units. The remaining three chassis were never sold in view of their unsatisfactory performance. Fortunately, the company's engine problems were solved when Austin's Leonard Lord agreed to supply them with the Sheerline's 4–litre, 6–

Rear view of the Jensen produced for Len Lord in 1952. The car's handling was much improved when the springs were positioned beneath the rear axle, instead of on top of it. Thankfully this historic Jensen has survived

cylinder engine that was also seeing service in Austin *and* Morris trucks. Jensen's new Six (Series PWA) was therefore announced in 1949 and remained available until 1952. Contacts between Austin and Jensen were strengthened when the West Bromwich firm produced the A40 Sports model of 1951. Designed by Jensen's Eric Neale it was, in effect, a scaled-down version of the company's own Interceptor of 1950, built on a modified Austin A70 chassis. Power was provided by a single-carburettored version of the Austin Sheerline engine.

Lord also asked Jensen to design a 2–seater sports car to replace the A40 Sports. As recounted earlier, Eric Neale produced a good-looking model but Lord preferred the Healey 100. Nevertheless, Jensen were awarded the bodywork contract for the new Austin Healey so this was adequate compensation. Production, by this time, was transferred to a new works in Kelvin Way, West Bromwich.

Not that all this work for Austin should detract from Jensen's own car-building activities. Their 541 model of 1954 used a predominantly glass-fibre body designed by Richard Jensen, with Eric Neale responsible for detail work.

A further Austin contract came with an order for bodies for the new Austin Gypsy of 1958, a cross-country vehicle intended to challenge the Land Rover. In 1959 Jensen were taken over by the Norcros Group and the following year the company signed a contract with Volvo for the painting and trimming of their distinctive sports coupé, the P1800. Shells were initially produced at Cowley but mainly at the Pressed Steel Company's Linwood works, though in 1964 all the work was transferred back to Volvo's Swedish factory.

Jensen's next car was the C-V8 of 1963, styled by Eric Neale and again in glass-fibre, though the pre-war theme of an American engine was revived with the fitting of a 5.9–litre Chrysler V8. Torqueflite automatic transmission by the same manufacturer was employed, power being taken through a limited slip differential. Jensen claimed another 'first' as the car was the first British production model to be fitted with an alternator. A Mark II version with a more powerful 6.2–litre engine appeared in 1964.

The company's next model was the Interceptor; styled by Touring of Milan it was the best-selling Jensen of the day. Although bodies were initially built in Italy by Vignale (Touring were on the point of closure), production was soon transferred to West Bromwich. The running gear was based on the latest C-V8 and, on its announcement in 1966, a 4-wheel-drive version, designated the FF, was also available. The system, built under Harry Ferguson patents, also incorporated a Dunlop Maxaret anti-skid braking system — yet another notable technical innovation for Jensen.

Unfortunately there were troubles ahead. From 1964 Jensen had assembled the Sunbeam Tiger sports car for the Rootes Group and later for Chrysler, the new owners. But the model was dropped in 1967 and, though Chrysler had plans for a replacement sports car, these never materialized. A GT version of the Hillman Imp, code-named Asp, was also considered, Jensen again assembling Pressed Steel body panels, but this idea suffered a similar fate. The real money came from

the big Healey, but this was dropped by BMH in 1967. From thereon the company had to rely on the low-production, high-price Interceptor, so Norcros sold out in 1968 to merchant bankers William Brandt. In 1970 Kjell Qvale, a wealthy car distributor from California acquired a majority holding in the company, his intention being to produce a Healey sports car with a Vauxhall engine. This seemed appropriate in view of Jensen and Healey's close association over the years and few could argue with Jensen's policy of producing a low-cost, high-production sports car for the American market to return them to commercial viability. However, in 1972 when the Jensen–Healey appeared, it was powered by a twin-cam Lotus 2-litre engine which, together with the bodywork, was beset with the kind of teething problems characteristic of under-development. The Jensen–Healey soon acquired a reputation for unreliability and the oil crisis sealed its fate. A receiver was appointed and production ceased in 1976.

Jensen production 1946–1967

$4\frac{1}{2}$–litre straight eight (HC series)		4
Straight Eight (PT and PW series)		3*
*produced but never sold		
Six (PWA series)		17
Interceptor		87
CV8	314	
CV8 Mark III	182*	
TOTAL	496	
*includes one prototype and one left hand drive car, both of which were sold		
FF (to 1971)		318
Interceptor (to 1976)		6387

541 1954-1958
541R 1958-1960
541S 1961-1963

Jensen 541S pictured on London's Westminster Bridge in September 1961. The model's appearance benefited from a conventional radiator grille in place of the flap previously employed on the 541 and 541R

Specifications and developmental history

The Jensen 541 was introduced at the 1953 Motor Show. A glass-fibre 2-door, four-seater saloon, with a frontal air intake flap controlled by the driver, was mounted on a chassis built up around two 4-inch diameter tubes re-inforced with box sections. Coil-and-wishbone independent front suspension was employed and Girling servo-assisted brakes were fitted. Steering was cam and roller. The 3993cc (87 × 111mm) engine was an Austin overhead-valve 6-cylinder unit, developing 130bhp. The interior featured bucket-type front seats and leather upholstery throughout. Price was £1771.

In October 1956 a de luxe version of the 541 was announced with a high-compression cylinder head, overdrive and tachometer. Dunlop disc brakes were fitted all round. The 541R appeared in October 1957. The engine, which developed 140bhp, was briefly a DS7 unit introduced on the Austin Princess IV in 1956 with twin SU carburettors instead of the triple ones fitted previously. Rack and pinion steering was employed and piston dampers replaced the lever type. The chassis tubes were increased to 5 inches diameter. Disc brakes were transferred to the R, the standard version briefly reverting to drums. The 541S was announced in October 1960 with a mildly re-styled body 4 inches wider and giving $1\frac{3}{8}$ inches greater headroom. A normal grille was fitted and a radiator blind replaced the frontal flap. The 135-bhp power unit continued triple SU layout and a Rolls-Royce manufactured automatic gearbox was standardized. An AC mechanical petrol pump replaced the electric unit. The facia was redesigned, a radio introduced as standard and seat belts were included in the specification. Production ceased in 1963.

Check points

Body and chassis

The 541's major advantage is its practically all-glass-fibre body with the exception of the aluminium doors. Therefore, though rust problems are not experienced, there are difficulties with the glass-fibre crazing. At the front of the car evidence of this can be found around the large air intake duct, while at the rear it appears around the flashing indicators. Oddly enough the glass-fibre is particularly thick on most 541s, though a few are surprisingly thin.

If the door handles are in a poor condition it is possible to obtain replacements as they are identical to those used on the Reliant Robin. On the basic 541 the boot hinges (which are chrome-plated Morris Minor bonnet hinges!) are at the bottom, though on the R and S variants they are situated on the top of the lid.

The Jensen's chassis is built around two sturdy tubes, the nearside one is sealed and serves as a vacuum reservoir for the brake servo. These tubes are not too troublesome, but the cross-member positioned above the rear axle is prone to rust at its extremities. Any problems here can be observed by removing the back seat squab and battery cover, which is situated on the offside. Alternatively, it can be examined by removing the spare wheel from its container beneath the boot.

Suspension, steering and brakes

Austin A60 independent front suspension units

are used on the 541 and it is worth mentioning that the model was built for cross-ply tyres rather than radials. If the latter are fitted the condition of the king pins should be checked as the covers put undue stress on them and they have been known to fracture if left unattended. Unfortunately, replacement pins are now difficult to obtain. At the rear semi-elliptic springs are fitted but the forward spring shackle should be examined, as the welding between it and the tube can deteriorate.

Initially the 541's steering box was a reversed, left-hand drive, Standard Vanguard unit, though today parts are difficult to locate. However, the R and S models were fitted with rack-and-pinion steering, which is also common to the MG Z series Magnette. Provided it is regularly maintained, this is usually trouble-free, though the rubber mounting bushes can wear causing the rack to 'float'. Fortunately, Austin 1800 bushes designed for a similar job can be substituted.

Brakes are reliable and the Dunlop disc system employed was also used on the Jaguar XK150. Inevitably this suffers from a mildly ineffective handbrake, but now the Jensen Owner's Club can supply a modification to correct the problem.

Engine, gearbox and transmission

The 541 used the Austin Sheerline/Princess 3993cc 6-cylinder engine in DS5 form, and can run on two-star petrol — a distinct advantage these days! For the R model, the first 57 examples used the twin-carburettored engine from the Princess IV DS7 which provides an extra 38bhp. Carburettors were also fitted on the offside, which was a departure from earlier practice. Unfortunately, the existing DS5 was then re-introduced and performance suffered.

This durable engine, which holds no less than 2 gallons of oil in its sump, is a very reliable unit and capable of well over 100,000 miles between overhauls. An acceptable oil pressure reading is 55psi at 1500rpm; anything less than this should be regarded with suspicion.

The Austin gearbox is robust, but the R type used a Moss all-synchromesh one, the Longbridge unit not offering this facility on bottom gear. There

are no particular difficulties with this or the Rolls-Royce automatic used on the S type. Also the Salisbury rear axle is very reliable.

Spares

Owing to the good offices of the Jensen Owners' Club, spares are in good supply. They hold all the moulds for body panels for these cars, which is a particular advantage if a replacement is needed. Also available are such items as wind-screens and the Perspex rear and side windows, which are otherwise difficult to obtain. As many of the mechanical parts were shared with the Austin Princess, it is worth noting that Classic Restorations of Brant Broughton, Lincs specialize in this particular model.

Jensen 541 spares

Classic Restorations, Midvil House, Brant Broughton, Near Lincoln LN5 OSH (tel: Loveden 72674).

Above, **the 541 engine. Triple SU carburettors are fitted to the 4-litre Austin engine of this 1955 car. Stout tubular supports are needed to support the bonnet in the raised position**

Left, **the 541S's interior. Much use has been made of padded safety rolls. Control lever for the Hydramatic automatic transmission is on the left of the steering column**

Jensen Owners' Club Ltd., Peter Wallis, Chalfont, 7 Ashley Road, Farnborough Hampshire.

541 performance and consumption, *The Motor*, September 14, 1955.
Top speed: 115.8mph. Overall mpg: 22.6 for 1148 miles.

Jensen 541 production

541	225
541R	493
541S	127*
TOTAL	845

*includes first two experimental cars, later sold

Jowett 1906–1953

Benjamin and William Jowett opened a small workshop in the Yorkshire town of Bradford at the turn of the century to undertake bicycle repairs. However, they soon began to dabble with the internal combustion engine and, in 1901, produced their own proprietary V twin, proudly initialled 'The Jowett Motor', and intended as a replacement for the contemporary de Dion and Aster units. This paved the way for the production of their own car which was a 7HP 2–seater, registered in February 1906. Output did not begin until 1910, the tough little 816cc model perpetuating the theme of the prototype's flat twin engine. From thereon all production Jowetts used this layout until output ceased in 1953.

In 1920, a new factory was completed at Idle, on the outskirts of Bradford, and production was transferred there. The Jowett car was made initially as a 2–seater, but in 1923 a 4–seater appeared, joined 2 years later by a lumbering saloon. Although Jowett experimented briefly with a conventionally-engined car in the mid-'30s, when eventually in 1936 a four-cylinder model appeared it was a 1166cc Ten with the same horizontally-opposed layout as the smaller model.

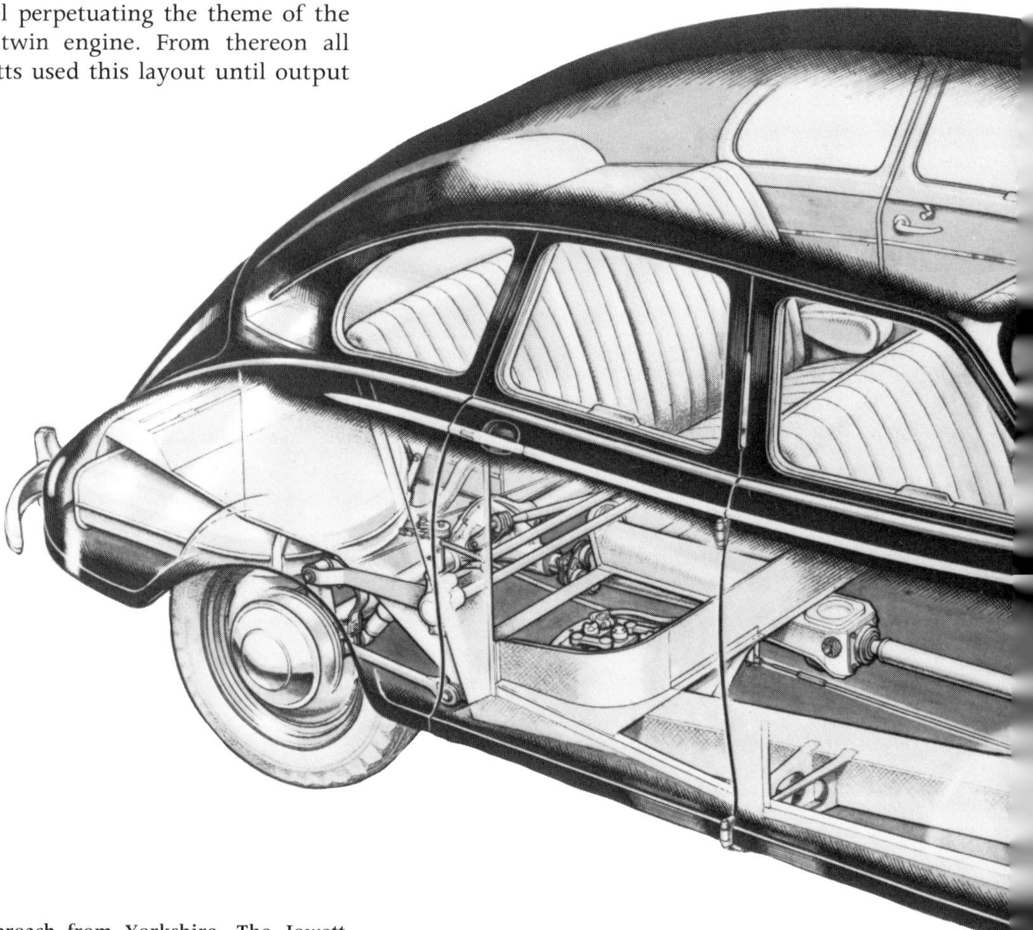

Unconventional approach from Yorkshire. The Jowett Javelin of 1947 was designed by Gerald Palmer and unlike most of its contemporaries was new from stem to stern. The forward mounted flat four 1½–litre engine was inspired by the pre-war German Steyr's layout and the styling by the American Lincoln Zephyr. Suspension was by torsion bar, longitudinal at the front and transverse at the rear. Unfortunately such sophistication was beyond the small Jowett Cars and production ceased in 1953

In 1935 family control diminished when Jowett became a public company and, in 1939, the energetic Callcott Reilly was appointed managing director.

During the war cooling pumps for Rolls–Royce Merlin engines and armour piercing shot and shell were produced, while hundreds of Sherman tanks were converted for mine-sweeping operations. Also the long-running, flat twin engine formed the basis of a self-contained generating set produced by the company. Reilly was also responsible for securing the services of Gerald Palmer to create a new Jowett for the postwar years and the design that he introduced was a thoroughly modern concept in stark contrast to the two 'lungers'.

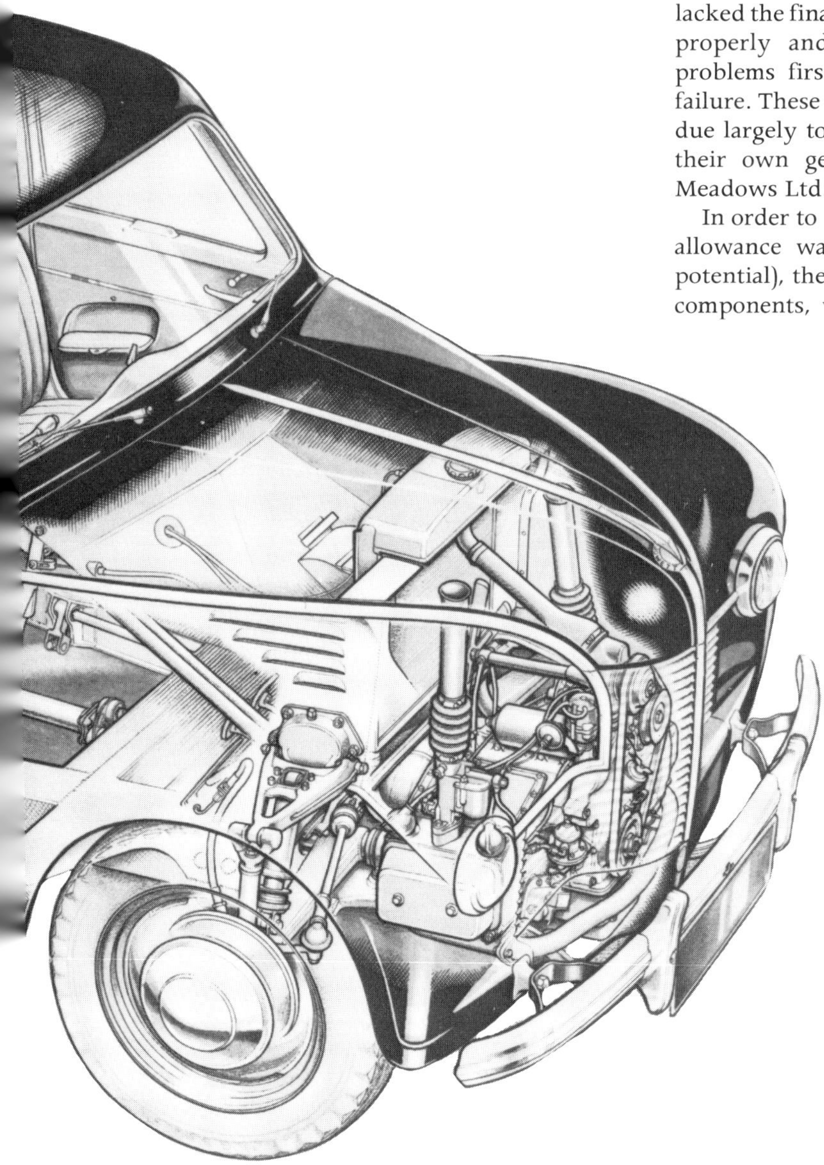

The Jowett Javelin prototype of 1944, had fashionable fastback styling, inspired by the American Lincoln Zephyr and the power was delivered by a 1.5–litre, horizontally-opposed, flat four engine, mounted well forward of the front wheel axis. Torsion bar suspension was employed all round. The Javelin didn't go into production until 1947, the bodies being supplied by Briggs Motor Bodies Ltd., who had a factory at Doncaster. However, in order to set the manufacturing wheels in motion, the utilitarian Bradford van went into production in 1946, perpetuating the 2–cylinder theme of pre-war days.

The Javelin soon proved to be an outstanding and lively performer, offering excellent passenger comfort and road holding. However, the company lacked the financial resources to develop the design properly and the engine particularly, caused problems first with gasket and then crankshaft failure. These were coupled with gearbox troubles, due largely to the company's decision to produce their own gearbox too hurriedly, after Henry Meadows Ltd. terminated the contract.

In order to boost their supplies of steel (tonnage allowance was assessed on the basis of export potential), the Jupiter sports car, based on Javelin components, was conceived. The tubular chassis

was designed by Eberan von Eberhorst, who had been responsible for the Grand Prix Auto Unions of 1937–1939, and the initial development was undertaken by ERA Ltd., at Dunstable. Production commenced at Idle in 1950, but the car did not prove a financial success, though it performed extremely well at Le Mans when it won the 1.5-litre class in 1950, 1951 and 1952.

Unfortunately, the Javelin's unreliability gradually damaged its reputation and demand plummeted drastically in 1952. Hulls were stockpiled at Bradford but, at the end of the year, Jowett asked Briggs, their body suppliers, to cease production. Early in 1953 matters were further complicated by the fact that Briggs, who had always built bodies for Ford at Dagenham, were taken over by them. Jowett did not repay their debts to Briggs until May, but they could not fulfil the terms for payment for the hulls of their new CD model. This employed a flat twin overhead inlet/side exhaust engine and was intended for manufacture initially as a replacement for the ageing Bradford and possibly as a saloon car.

So production never re-started and all remaining Javelin bodies, some of which had spent many months exposed to the elements, were finished off at Idle during 1953. The R4 Jupiter sports car, with

Above, **the Jowett Jupiter with chassis by Eberan von Eberhorst and styling by Reg Korner. Mechanics were Javelin based. A 1952 Mark 1 A version is shown, improved by an opening boot. Radio was an optional extra.** *Right*, **a 1952 Javelin de Luxe, showing the lines to good effect**

its partial glass-fibre bodywork, appeared at the Motor Show 1953. It was conceived by Roy Lunn, who had taken over from Gerald Palmer in 1949, but the car signified the end, rather than the dawning of an era. Lazard Brothers, the London merchant bank who had controlled Jowett's finances since 1947, decided to call it a day and the Idle factory was sold to International Harvester in 1954.

When the Bradford van also ceased in 1953, its flat twin engine, by then in 1005cc form, had been in continuous production for 43 years, a notable British record.

Jowett production 1946–1953

Bradford	38,241*
Jupiter	approx. 850
R4 Jupiter	3

* Body production figures

Javelin 1947–1953

Specifications and developmental history

The Jowett Javelin was first fully described in the motoring press in May 1947, and went into production later in the year. A 4–door, 5–seater saloon, it employed an A frame box-section chassis, joined at the rear by a sheet steel assembly. Suspension was by wishbone-and-torsion bar at the front and transverse torsion bar and radius arms to a live rear axle. Brakes were Girling Hydro-mechanical (hydraulic front and mechanical rear) and a quadrant-and-pinion steering box was fitted. The engine was an overhead valve, flat four of 1486cc (72 × 90mm) developing 50bhp. Leather or leather-cloth upholstery was available. The car cost £818.

For 1950 Standard and De Luxe versions appeared, the latter with a wooden facia with round-faced, recessed instruments, armrests for each passenger and a heater as standard; re-designed bumpers were fitted. Cars for 1951 featured separate side lights instead of their being contained in the headlamps, while at the rear, De Luxe examples were given a bumper mounted number plate box. Major mechanical changes consisted of: conventional tappets replacing the Zero-Lash hydraulic ones; full hydraulic brakes and the fitting of an Idle built gearbox (from car no. EO/PC/11270) in the spring of 1951. Further changes at the beginning of 1951 included oil pressure increased from 60psi to 70psi, the oil filter being changed from a Vokes to a full flow Tecalemit. The radiator grille was also re-designed. Javelins for the 1952 season were fitted with Metalastik independent front suspension bushing. The De Luxe wooden facia was re-designed with flush-fitting instruments with chrome bezels. The more reliable Series III engine was fitted in cars in 1953, which was the final year of production.

Check points

Body and chassis

The Javelin's box-section chassis stops well short of the rear of the car and this is the first point of investigation. Underneath the car any suspect areas should be tested by prodding with a screwdriver. The forward end of the rear wheel arch is a particularly vulnerable section for two reasons. Apart from being blasted by grit and mud thrown up from the road, the location is also under stress as the rear transverse torsion bars are mounted there. The chassis should be checked at the same time. Rust may have attacked the front and rear outriggers. The latter also serves as a jacking point and is found immediately beneath the central door pillar, which in extreme cases can also suffer. The front chassis cross-member should also be examined for splits along the seam (it was a favourite spot for a careless mechanic to place a trolley jack and the box section does not take the strain).

The bodywork doesn't seem to suffer too much at the front, but at the rear the floor of the boot should be checked as it is particularly susceptible to rusting. And on the De Luxe cars the steel surround containing the spare wheel is also likely to disintegrate. It is, undoubtedly, the rear of the Javelin that demands the most careful inspection.

A final word of warning: particular care should be taken when examining a Javelin registered in late 1952 or 1953. The sales dive of the previous year and resultant stockpiling of bodies outside the Idle factory meant that many of the shells were already starting to rust when they were assembled.

A few words are probably worthwhile on the Jowett numbering system. A typical Javelin car number might be D9/PA/10631. The first letter indicates the decade, A is 1, B 2 and so on. The second digit indicates the years of the decade. The third letter P indicates a 'pleasure model', as opposed to C for commercial. The fourth letter shows the basic model type, which changed as modifications were introduced. The final figure is the engine and chassis number. The example given shows that the car is a 1949 pleasure A model (early type) Javelin. As it happened the model year was changed every year from 1949 to 1953. Therefore a late model 1953 car would be E3/PE/28562.

Suspension, steering and brakes

Suspension faults are usually confined to the Javelin's front end. The main difficulty is that the upper wishbones pivot on single bolts and if they loosen the resultant movement can crack the chassis. To avoid this the Jowett Car Club recommend that a high-tensile Allan-headed *Unbreako* bolt is fitted in place of the original. The upper wishbone brackets also act as oil reservoirs, which supply lubricant through a tube to the swivel pin yoke. This method of lubrication was discontinued after car number D9/PB/5979 and a grease-nipple at the upper end of each upper link substituted. The reservoirs require topping up with 40 grade oil every 500 miles and excessive wear of the swivel pin yoke indicates negligence in this regard. The system was replaced in 1952 when Metalastik joints were introduced. The rear torsion bar layout is comparatively trouble-free, apart from rust of the torsion bar mountings already mentioned.

Steering gear requires lubrication every 2500 miles though the joints need greasing at 500-mile intervals so excessive wear due to lack of maintenance is not uncommon.

There aren't any special problems associated with the brakes and obviously the full hydraulic layout fitted from 1951 models onwards was a great improvement to the Hydro-mechanical system employed earlier.

Engine, gearbox and transmission

Although Jowett did have their fair share of mechanical misfortunes with the unusual flat four engine fitted in the Javelin, most examples have by now been rectified. The Series III engine of 1953 was the most reliable unit of all, but the much improved oval webbed crankshaft, designed for the company by Dr Ker Wilson of De Havillands, was not fitted to reconditioned engines until about 1955, 2 years after car production ceased.

Nowadays, the main disadvantage with this engine is the aluminium block. Over a long period, the water within the block and the weather conditions cause severe corrosion to the underside and repairs are tricky and expensive. If possible, the bottom of the block should be examined for signs of damage or bodged repairs. Also the cast-iron cylinder heads tend to crack between the valve seats and a damaged gasket probably means that the engine's wet liners weren't 0.008–0.010 proud of the block when the engine was re-assembled after a de-coke. A torque wrench is a vital tool when working on this engine because of the amount of aluminium it contains. The distributor is rather exposed to the elements and misfiring can be prevented by wrapping it in a polythene bag. If the sparking plugs have to be removed, the plug wells should be free from stones and grit otherwise they will drop into the combustion chambers. An acceptable oil pressure reading for this engine is 50psi at 40mph.

The Javelin was fitted with two different gearboxes during its production life. Up until the spring of 1951 a Meadows-built unit was employed. This tends to be fairly reliable, though it can jump out of third gear. From thereon a Jowett-built box was fitted and this, amongst other shortcomings, can strip the teeth off its second gear. Although in contemporary terms this gearbox was less reliable than the Meadows one, many examples have continued to operate reliably for years. The steering column gear change is a well-engineered component and shouldn't develop any slack.

Power was transmitted to the rear axle via a two-piece propeller shaft supported centrally by a bearing which in turn was contained by a cross-member and held in place by fabric or rubber strips. The condition of this supporting material is vital and should be examined for chafing or wear, which could cause bad vibrations. The rear axle is not prone to trouble.

Spares

Like most saloons of its age, it is virtually impossible to obtain replacement body panels for the Javelin. However, so far as mechanical spares are concerned the Jowett Car Club has recently taken over a large stock, while another source is Dennis Sparrow and Associates of London who also undertakes work on these rather unconventional cars.

Javelin spares and service

Dennis Sparrow and Associates, 77 Jeddo Road, London W12. (tel: 01-749 2013).

Jowett Car Club, Dr Harry Brierley, 5, Farne Avenue, Gosforth, Newcastle-upon-Tyne.

Javelin performance and consumption (Series III), *The Motor* April 8, 1953. Top speed: 82.4mph. Overall mpg: 29.1 for 2290 miles.

Javelin production 22,799*

*body production figure

Right, a bench type front seat was possible by the use of a steering column gear change. _Below_, open wide! Some of the Javelin's engine revealed though sparking plugs were rather more difficult to get at

Lotus 1953 to present date

Anthony Colin Bruce Chapman was born in 1928 in Richmond, Surrey but 2 years later his father took over the ownership of the Railway Hotel, Hornsey so the family moved to London. In 1945 he entered University College, London where he met up with Colin Dare. They went into part-time business buying and selling cars as there was a demand for *anything* on wheels in the car-hungry postwar years. Unfortunately for them their activities came to an abrupt end when the basic petrol ration was cancelled in October 1947. One of the cars left on Chapman's hands was a 1930 Austin Seven fabric saloon, registered PK 3493, and he decided to scrap the body and convert it into a Special. No substantial modifications were made to the engine and suspension at this stage, but the well-executed, open, doorless body was a sign of things to come. It was of alloy-bonded plywood mounted on a stressed framework structure with triple bulkheads, designed on aircraft principles to re-inforce the Seven's whippy A frame. The really important thing about this car is that, after it was completed in 1948 it was registered OX 9292 as a 'Lotus'.

In 1948 Chapman left university with a B.Sc. in engineering and joined the RAF to do his National Service. Afterwards, in 1950, he went to the British Aluminium Company as a construction engineer, though he continued to produce a trickle of one-off specials and sports cars. Then with a fellow enthusiast, Michael Allen, he formed the Lotus Engineering Company, which went into business in stables at the rear of the elder Chapman's hotel, to set production on a more organized footing. This was in January 1952 and the Lotus Mark VI was created in these somewhat inauspicious surroundings, the first of Chapman's models that didn't employ an Austin Seven chassis. Unfortunately, the prototype was written off in an accident and Allen departed taking the badly-damaged car with him as his share of the assets! Chapman, therefore, borrowed £25 from his girl friend, Hazel, added £100 of his own and the Lotus Engineering Company *Limited* was re-formed in February 1953. He was then in a position to offer the Mark VI for public sale, though it should be added that the business was very much of a part-time affair as Chapman was still working full time for British Aluminium. The club racer was sold mostly in kit form having a multi-tubular chassis with coil spring and swing-axle independent front suspension while a live axle and coil springs were fitted at the rear. Owners used a variety of engines ranging from 1172cc side-valve Ford to a 2–litre BMW unit. This lasted until 1955, a year in which Chapman devoted himself exclusively to Lotus activities. Team Lotus, which was solely concerned with racing, had been established the previous year. The company produced sports racing, later formula 2 and finally formula 1 cars, the marque having been in the forefront of Grand Prix racing since 1960.

However, our concern is Chapman's production of road cars. The first significant example was the Lotus 14 — the Elite — which made its debut at the Motor Show in 1957. Styled by Chapman's friend, Peter Kirwan-Taylor, the Elite was a 2–door, glass-fibre, monocoque GT coupé powered by a 1216cc, single overhead camshaft, FWE Coventry Climax engine. Suspension was by coil-and-wishbone at the front. At the rear was a long Chapman strut, a combined coil/spring damper layout, with wheels located by drive shafts and a kinked single radius arm. It was not until December 1958, however, that the first production Elite was handed over to its owner, bandleader Chris Barber.

Chapman was rapidly outgrowing his Tottenham Lane premises and in June 1959 Lotus moved to Delamare Road, Cheshunt, Hertfordshire, where Elite production really got under way. A Series 2 version with revised rear suspension, a more powerful engine option and a ZF all-synchromesh gearbox appeared for 1961. The following year the Elite was available in kit form, which was well-established Chapman practice. Output ceased in 1963. Produced alongside the Elite was the Lotus Seven, a rather basic 2–seater with outstanding road holding but with weather equipment that was not for the fainthearted!

A new model, the Lotus type 26—the Elan—was launched for 1963. A shortcoming of the Elite was that it could not be made in open form because the roof was an integral and vital part of the monocoque structure. Therefore, though it employed glass-fibre bodywork, the 2–door, 2–seater Elan used a new fabricated steel backbone chassis with the opposite ends splayed out to receive the engine and rear suspension. The power unit was 1498cc twin overhead camshaft and Ford based, development work having started in 1961. Harry

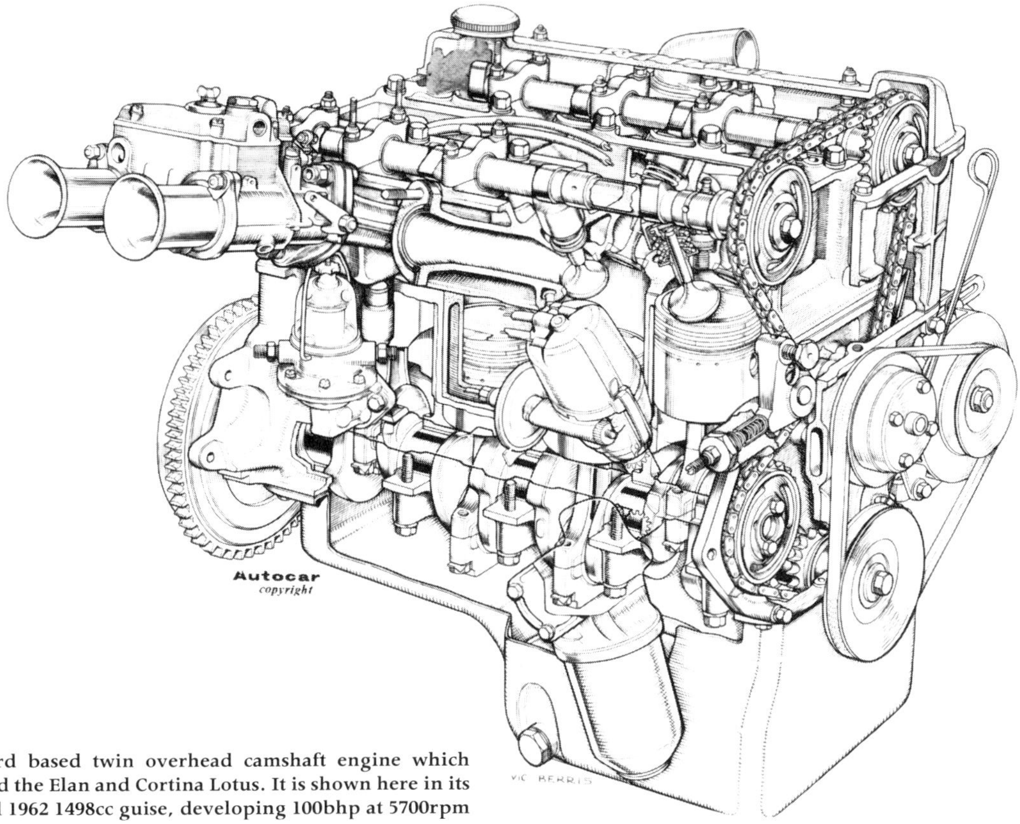

The Ford based twin overhead camshaft engine which powered the Elan and Cortina Lotus. It is shown here in its original 1962 1498cc guise, developing 100bhp at 5700rpm

Mundy, formerly of Coventry Climax and at the time technical editor of *The Autocar*, designed an alloy twin-cam head, which was initially fitted to the 109E block. However, the appearance of the 116E five-bearing crankshaft engine, destined, briefly, for the Ford Classic and for the new Cortina of 1962, meant that production engines for the Elan were based on this block. Assembly was assigned to J. A. Prestwich, famed for their JAPV–twins of pre-war days, and after about 50 Elans had been built, the capacity was increased to 1588cc and the earlier cars recalled. This co-operation between Lotus, Cosworth Engineering, who had been responsible for developing the engine, and the Ford Motor Company was to strengthen during the decade. Yet another instance of the alliance was the appearance of the Cortina Lotus (type 28) where the Elan engine was fitted in a Mark I Cortina saloon at Cheshunt, as chronicled on page 144.

Development work on the Elan continued apace. The Series 2, which appeared in November 1964, had a better finished interior and larger front brakes. A coupé arrived late in 1965 (type 36), while the following year the open version was discontinued in favour of a drophead coupé model (type 45). In mid-1967 the Elan +2 (type 50) appeared, which was a Lotus for the family man, having two rear seats for children. Production finally ceased in 1973.

The expanding Lotus company had again outgrown their premises and moved to Hethel, near Norwich at the end of 1966. These later cars were built at the new factory, and the twin-cam assembly was transferred there at a later date. The move coincided with the announcement of a new road car, the Europa (type 46)—the first sensibly priced mid-engined car to go into production anywhere in the world. It used the 1470cc engine from the Renault R16 saloon, though with a higher compression ratio and special camshaft, and was initially only available for the export market. Production started in 1967 when Chapman was employing a workforce of over 500 at Hethel. Those days at Hornsey must have seemed a long way off. . . .

Lotus production 1959–1967

Elite	998*
Europa (to 1975)	9905

* several more examples probably built up from parts

Elan 1963-1973
Elan +2 1967-1973

Specifications and developmental history

The Lotus Elan was announced at the 1962 Motor Show. A 2-door, glass-fibre, open 2-seater body, with distinctive pneumatic operated headlamps, was mounted on a steel backbone chassis. Suspension was independent all round; wishbone and coil spring at the front with Chapman struts and single lower wishbones at the rear. The engine was a twin overhead camshaft Ford 116E-based unit of 1498cc (80 × 72mm) developing 100bhp. The price was £1499, or £1095 in kit form.

In May 1963 the engine size was increased to 1588cc (82 × 72mm) developing 105bhp (from chassis 26/0026). 'Elan 1600' flashes fitted on front wings and hardtop became an optional extra. The Series 2 Elan was introduced in November 1964 (at chassis 26/3901). Larger front brake calipers were fitted and individual rear lights merged into an oval unit. 'S2' motif was added. Interior improvements included a full-width veneer dashboard with lockable glove box and chrome bezel instruments. In September 1965 the Series 3 Elan appeared (36/4510) as a fixed-head coupé with electric windows. The battery was relegated to the boot. High axle ratio of 3.55 was available. November saw the arrival of a close ratio gearbox (36/147). A Special Equipment model became available in January 1966. Engine output increased to 115bhp with close ratio gearbox, servo-assisted brakes and centre lock wheels standardized. Repeater flashers were fitted to front wheel arches. A Series 3 version of the convertible appeared in June 1966 (45/5702).

In 1967 the Elan +2 fixed-head coupé was introduced with engine power increased to 118bhp (50/0001). Construction was similar to the Elan but 2 feet longer and 10 inches wider, to provide two small rear seats for children. Throughflow ventilation was fitted. The Series 4 arrived in March 1968, in coupé and drophead coupé forms. Bulge on the bonnet is an identifying feature and flared wheel arches indicate low-profile tyres. The facia was re-styled with rocker switches and perforated seat trims were introduced (45/7895).

The Elan 1600 of 1964 with headlamps in the operative position. The glass fibre bumper was filled with foam rubber

The Elan + 2S was announced in October 1968 but did not go into production until March 1969 and was the first Elan unavailable in kit form. Specifications were similar to Elan + 2 but with improved interior. Fog lamps were standardized and the '+2' emblem appeared on the boot (50/1554). From November 1968 Stromberg carburettors replaced Webers on all models except the Elan +2S (50/1280), on +2 fixed-head coupé (45/8600) and on the convertible and fixed-head coupé Elans. However, in August 1969 Weber carburettors were re-introduced on the convertible (45/9524) and fixed-head coupé (36/9524). The Elan + 2 was discontinued in December 1969, though the +2S continued in production. In February 1971 a Sprint version of the Series 4 was introduced with a more powerful 126bhp engine and a strengthened differential, drive shaft and drive couplings. It was identifiable by duotone paintwork. At the same time the Elan +2S 130 appeared, sharing the same engine as the Sprint and distinguishable by its Silver roof. For 1973 a 5–speed gearbox was introduced on the Elan + 2S 130 denoted by the number S130/5, (at chassis 7201). In August 1973 all models ceased production.

Note: Chassis numbering changed from January 1970 from number 7001 010001. Models were continued with an identifying suffix, rather than a prefix. Therefore Series 4 fixed-head coupé = A, special equipment = E; convertible = C, special equipment = G; + 2S fixed-head coupé = L.

Check points

Body and chassis

Fortunately there are no rust problems with the Elan's glass-fibre bodywork, though star cracks, particularly along the lower surface of the material, are often caused by stones thrown up from the road. It is, of course, quite possible for replacement parts to be fitted and, if there is a ridge on the underside of the panels, this probably indicates that a new section has been fitted. Perhaps the Elan's most distinctive feature is the pop-up headlights. Initially they were kept in their operative position by manifold pressure but on the + 2S 130, the lamps became fail-safe with the vacuum being responsible for keeping the lights *down*. Don't be surprised if the lamps wobble — they all do! Difficulties in this area are usually caused by splits

The Elan +2, announced in 1967, almost 2 feet longer, 1 foot wider and 1 inch higher than the Elan, pictured at Lotus's new factory at Hethel, Norfolk

in the rubber piping which convey the 'vacuum', faulty unions or troubles at the non-return valve.

The distinctive backbone chassis is well protected and shouldn't suffer from rusting problems. As a guide, check the state of the chassis member in front of the engine and if it has succumbed to bad corrosion, replacement will require major surgery. On the Elans and + 2, this member also served as a vacuum reservoir for the pop-up headlights.

Suspension, steering and brakes

Wear in the front suspension is usually caused by worn trunnion bushes. (The trunnion is the ball-joint housing attached to the wishbone on which the upright pivots.) The bush is located clamped between the outer ends of the lower two-piece wishbone. It is also important that the bolt running through it is tight. If it isn't then the hole in the wishbone can become elongated which isn't very desirable. If the car responds badly at the front end look there first. Bent steering rack mountings can produce similar symptoms often caused by the car having been carelessly jacked up. As a guide, the track rod arms should adopt equal angles on level ground. At the back of the car check first the chassis end of the torque bars locating the lower part of the differential. These bars *must* be taut on their bushes and, if there is any fore and aft movement, then a cracked chassis can result. The hardworking dampers, that should also be checked for leaks, play an important role in the model's suspension system. When the rear of the car's bodywork is depressed manually and then released it should just return to its static position. A very small amount of additional movement is permissible, but no more. Another likely problem is a hardening with age of the rubber top mountings of the rear dampers and any noisy suspension 'bonks' audible on the test drive are probably attributable to this. Worn bearings can be detected but it should be remembered that slack might be caused by wear in the lower wishbones' rubber bushes.

Brakes shouldn't present any major problems. Many + 2s were fitted with protection shields for the rear brakes, but these weren't fitted to the short wheelbase Elan. Neither were they employed on the earlier + 2s, but they are worth mentioning because, if they are used, they make the inspection of the outer drive shaft doughnuts rather more difficult when you come to check their condition.

Engine, gearbox and transmission

Fortunately, the twin-cam engine is a reliable unit but an oil consumption of 1 pint per 200–400 miles is not surprising. The early engines built by J. A.

Well placed controls. The facia is wooden and the instruments well positioned on this 1964 car. *Right*, the bonnet was completely removable. Sparking plugs are concealed by a metal cover

Prestwich are identifiable by the steel tappets running directly in the aluminium cylinder head, though an early modification was sleeves for the exhaust buckets. You can confirm the presence of the sleeves by removing the camboxes and checking at the base of the valve springs. It should be noted, however, that many Mark I engines were converted to this specification. The Mark II version appeared in 1967; by this time Lotus were producing the engines themselves (from engine number LP 1800) and these examples have both the inlet and exhaust buckets sleeved. Other changes were stronger interchangeable conrods marked 125E, pistons with a smaller crown in place of high crown ones with bigger valve cut outs and a flywheel secured by six bolts rather than four.

It is important that these engines are oil-tight and, therefore, that the one-piece cork cambox gasket is in good condition. Incidentally, it is quite usual for the carburettors to wobble up and down when the engine is running. They are mounted on rubber O rings and this applies to all the makes of carburettor employed, namely Weber, Dellorto or Stromberg. Beneath the carbs is the distributor and ideally this must be removed to ensure correct adjustment. One component that will probably give trouble is the water pump. It tends to wear and if movement at the end of the fan blades (or boss if a plastic fan is fitted) is greater than 0.25 inches then the unit should be replaced. A good indicator of the state of the engine is the condition of the timing chain. It requires adjustment every 5000 miles and if it rattles this confirms that the chain needs attention. On the other hand a whirring sound indicates too much tension. The chain's adjusting nut can be found at the front offside of the engine about half way down and the prominence of the threads is an indicator of the chain's state. If there is about 0.5 inches exposed then you can reckon that the chain is in good condition and has plenty of wear left in it. But if the threads have been completely taken up then a new one is probably required. An acceptable oil pressure reading is 40psi at 3000rpm when the engine is hot. There is no cause for alarm if the engine smokes a bit under power as this is quite normal.

The usual gearbox was from the Corsair 2000E. If it jumps out of gear this is usually due to worn selectors. Some early units employed Lotus's own ratios in a Ford Classic casing. First series Elans suffered from a spot of gearbox zizz and this also applied to early Series 2 cars but this was cured by the introduction of a judiciously-placed Silentbloc bush with the top part of the gear lever welded to the outer ring.

The differential is reputed for its reliability, but the rubber drive shaft couplings at both ends of the shaft should be inspected for splits, which are usually adjacent to the bolt holes.

Spares

Fortunately, Lotus Cars still supply spares, both body and mechanical, for their earlier models. Ian Walker of Leighton Buzzard are official Lotus reconditioners of these twin-cam engines while, at the other end of the country Vegantune of Spalding, Lincs also offer a twin-cam re-building service. Club Lotus, with its regular programme of events, is certainly worth joining if you own one of these cars.

Elan spares and services

Lotus Group of Companies, Wymondham, Norfolk (tel: 0953 603411).
Vegantune, Cradge Bank, Spalding, Lincolnshire (tel: 0775 4846).
Ian Walker, Shilton Trading Estate, Leighton Buzzard, Beds. (tel: Leighton Buzzard 375851)

Club Lotus, Margaret Richards, 107 Brandon Road, Watton, Thetford, Norfolk.

Elan performance and consumption, *Motor*, September 26 1964.

Top speed: 111.9mph. Overall mpg: 25.5 for 2228 miles.

Lotus Elan production

1963		300 *
1964		1195 *
1965		960
1966		1387
1967		1382
1968		887
1969		1151
1970		970
1971		766
1972		543
1973		118
	TOTAL	9659
Lotus Elan + 2 production		4798

* estimates

M.G. 1924 to present date

M.G. stands for Morris Garages and wheels were set in motion for the marque's creation in 1922 when general manager Cecil Kimber ordered Morris Cowley chassis and had two–seaters rebodied in a more individual style. Production, such as it was, transferred from Longwall Street to a small workshop in Alfred Lane, Oxford early in 1923. Later Raworth-bodied two–seater Cowleys were produced and demand subsequently increased so that the cramped premises were replaced in 1925 by larger cramped premises in Bainton Road, which was part of the Morris Motors radiator factory. Kimber, no doubt becoming somewhat disgruntled at the constant moving, persuaded William Morris to sanction a purpose-built factory at Edmund Road and production was transferred there in 1927. These early MGs were side-valvers, 1802cc 14/28 or 14/40 models, but for 1929 came two new overhead camshaft cars, the 2468cc 18/80 and perky little 847cc M type.

Space again became a problem and the MG Car Company, as it had become in 1928, moved in the autumn of the following year to a separate factory at Abingdon, six miles from Oxford. There, at a works previously occupied by the Pavlova Leather Company, Kimber resolved to expand, particularly in the field of competition. Although the M type sold 3235 examples, this promise was not maintained, the 1271cc F type Magna being a less successful 6–cylinder model. However the J1 of 1932 and later J2 fared better, the latter model with its double-humped scuttle, cutaway doors and flat back body with external petrol tank set the pace for MG styling right up until the TF ceased production in 1955. These were followed by PA and PB fours, though by the time the latter version went into production in 1936 the engine size had increased to 939cc. The 6–cylinder cars continued with the L–type Magna, together with a variety of racing models produced in small numbers, of which the K3 of 1932/34 is perhaps the best known. MGs were actively campaigned in competitions at home and abroad, the marque's most famous victory coming in the 1933 TT when the great Nuvolari took the chequered flag at the wheel of a K3 a minute ahead of another Abingdon product.

But despite such successes, sales dropped and diversification increased. All models from 1931 used a sophisticated overhead camshaft design whereby the vertically mounted dynamo armature also served as the camshaft drive. Hubert Charles, Cecil Kimber's talented designer, produced a finely engineered range of cars but maintenance was difficult and performance easily upset. However, in July 1935 Lord Nuffield sold the MG Car Company, along with Wolseley Motors to his vast Morris Motors combine. The racing programme was abandoned and overhead camshaft engines replaced by Morris and Wolseley-derived pushrod units, while the Abingdon design office was closed down and transferred to Cowley. The outcome of all this upheaval was the 6–cylinder, 2.7–litre SA saloon of 1936 and later the 1.5–litre 4–cylinder VA, which helped to put Abingdon's products on a healthier commercial footing. In mid-1936 came the 2–seater TA Midget powered by a 1292cc pushrod engine and destined to be the best-selling MG since the original M type of 1929. The TB model that succeeded it in 1939 used a new XPAG engine based on the Morris 10, while a saloon version, the Y type, with independent front suspension was destined to appear at the 1939 Motor Show.

During the war MG undertook the production of a complex section of the Albemarle bomber, but it was a contract that led to Cecil Kimber's dismissal in 1941. Miles Thomas, the vice chairman and managing director of the Nuffield Organisation wanted the group geared to military projects and he clashed with Kimber when he heard of the Albemarle job. Kimber later joined Specialloid Pistons as works' director, but he was killed in a train accident near King's Cross Station in February 1945 at 59 years of age.

At the end of the War in 1945 MG went straight into production with a single model, the TC 2–seater — a mildly updated TB. The car sold very well by pre-war standards, one-third of the output being exported. Although initially sold in small numbers in USA, its impact was enormous and it spearheaded British sports-car sales in that continent. In 1947 came the Y type saloon, shelved since 1939, which remained in production in YB form until 1953. The independently sprung TD with disc wheels replaced the TC in 1950 and was based on a shortened form of the Y type chassis. It sold three times as well as the TC, but the time was clearly ripe for a new model, the MGs from Abingdon, appearing rather quaint in comparison with many of their contemporaries with their

Top, the export only YT tourer, announced in October 1948. A twin carburettored engine was fitted. *Above*, a batch of YA's leaving Abingdon for export to Malta

full-width streamlined bodies.

Things were complicated by the creation of BMC in 1952 and, as we have seen, Leonard Lord asked MG, amongst others, to design a sports car for BMC to manufacture. MG's Syd Enever produced a stylish 2–seater, based on a special bodied TD that ran at Le Mans in 1951, but Lord approved the Austin Healey 100 and EX175 of 1952 — the TD replacement — was sidelined. A stop-gap was the TF of 1954–55, latterly with a 1466cc engine.

In the meantime the Magnette name had been revived for the ZA of 1954, which used an existing Wolseley 4/44 body shell but was fitted with a new 1489cc version of the freshly termed BMC B series engine of Austin design. It was updated as the ZB for 1957 but production ceased in 1958.

Fortunately, 1954 saw the re-opening of the Abingdon design office, with John Thornley, who had been appointed MGs general manager in 1952, reviving the EX175 project which blossomed into the MGA of 1955. The model made its debut at Le Mans in 1955 in EX182 guise, the A going into production later in the year. Under the bonnet was the 1489cc B series engine, which this was increased to 1588cc in 1959 and again to 1622cc in 1961. The A was destined to become a best-seller, over 100,000 being produced before the model ceased in 1962. This included a limited run of twin overhead camshaft examples of questionable reliability built between 1958 and 1960.

The MG Midget, a badge-engineered version of

The MGA coupé in 1500cc form sold for £1049 in 1956. The export orientation of this model is readily apparent. Wire wheels were an optional extra

the Austin Healey Sprite made at Abingdon since its inception in 1958, appeared in 1961 (Series G/AN), though the capacity was later increased to 1098cc (G/AN3) in 1964. It was again enlarged to 1275cc for 1967 (G/AN4) and remained in production (though with a Triumph engine from 1974 until December 1979) long after the Sprite, from which it derived, ceased to be manufactured.

In 1962 came a replacement for the MGA and, unlike its predecessor that was built up around a sturdy chassis, the MGB initially available as an open two–seater (G/HN3) favoured a monocoque hull. The B series BMC engine was again used, though enlarged to 1798cc. A five-bearing crankshaft version appeared for 1965, while a coupé GT, designated model (G/HD3) went on sale for 1966. Perhaps the most radical alteration to the B's appearance came in the 1975 model with the fitting of flexible rubber bumpers front and rear, in deference to American safety regulations.

But what of the MG saloons? The Magnette had received Farina-styled coachwork in common with other BMC marques in 1959 to become the Mark III (GHS 1), while the 1622cc Mark IV (GHS 2) of 1961 remained available until 1968. An MG version of the top-selling Morris 1100 front-wheel drive car appeared for 1963 (G/AS); it also extended to the 1300 (G/AS4) model of 1967. But these cars were built at Cowley.

The other new MG model for that year was a 3–litre version of the MGB. The MGC was available in open (G/CN1) and coupé (G/CD1) form, sharing a 6–cylinder engine with the Austin 3–litre saloon that was announced at the same time, but that did not go into production until summer of 1968. The B's coil-and-wishbone independent front suspension was dispensed with and replaced by torsion bars, but this controversial model was only made up until 1969.

Although it is outside our period it is worth mentioning that, in 1969, Abingdon came up with a sensational mid engined design, a wedge-shaped car styled at Longbridge by Harris Mann and coded ADO21. It was conceived around an Austin Maxi 1750HL power unit, mounted transversely, though 2.2 or 2.6 sixes on the same theme were also considered. ADO21 remains the most tantalizing MG that never materialized because a conventionally-engined Triumph project was chosen instead and the outcome was the TR7 of 1975.

In March 1980 it was revealed that BL Cars (British Leyland's new title) had agreed to sell the Abingdon factory and granted an exclusive world licence for the MG name to a consortium headed by Aston Martin Lagonda. Unfortunately, on July

1st, BL announced that the consortium had been unable to raise sufficient funds. Therefore MGB has ceased production and the world famous Abingdon factory closed.

MG production 1945–1967

TC Midget	10,000
TD Midget	29,664
TF Midget (1250cc)	6200
TF Midget (1500cc)	3400
ZA Magnette	12,754
ZB Magnette	23,846
Magnette Mark III (Cowley built)	15,676
Midget Mark I	16,080
Midget Mark I (1098cc)	9601
MGB Mark I	115,898
MG 1100 (Cowley built)	134,680
Midget Mark II	26,601
MGB GT Mark I	21,835
Midget Mark III (1967 Cowley built)	
Midget Mark III (to 1968)	13,246
MGC two-seater (to 1969)	4542
MGC GT (to 1969)	4457

$1\frac{1}{4}$-litre, series Y 1947-1951 series YB 1951-1953

Specifications and developmental history

The MG $1\frac{1}{4}$-litre was announced in May 1947. It was a four-door saloon, with sunshine roof, mounted on a box-section chassis with coil-and-wishbone independent front suspension, Lockheed hydraulic drum brakes and rack-and-pinion steering. The engine, a Morris Ten-based XPAG unit, was a pushrod overhead valve four of 1250cc (66 × 90mm) developing 46bhp. A noteworthy feature was a built-in *Jackall* hydraulic jacking system actuated manually from under the bonnet. Price was £672.

A touring version, the YT, with 4-seater open body and 54bhp twin carburettored engine ap-

The Y series saloon sold fairly well on the British market and overseas though this 1951 example's pre-war origins are all too obvious

peared for the 1949 season but production ceased in 1951.

The improved Y B saloon of 1952, identifiable by the deeper valancing on the rear wings, incorporated many components from the TD 2–seater, with 15–inch wheels in place of the Y A's 16–inch ones, twin leading shoe brakes, hypoid rear axle and stronger shock absorbers. A front-mounted anti-roll bar was also introduced. Production ceased in 1953.

Check points

Body and chassis

The Y type's body is an all-steel affair and it suffers from rust around the rear end. The most serious shortcoming is that the rear body mountings corrode and, in extreme cases, the boot can be heard banging on the chassis when the car is in motion. The condition of these mountings can be assessed by inspecting inside the spare wheel locker to be found under the boot. If they are badly affected then repair can be difficult and expensive. Another set of body mounting bolts are situated inside the car beneath the carpeting in front of the rear seats. Check also around the inside of the rear wheel arches as this is where the drain rubbers from the sunshine roof terminate. It is important that these tubes are clear and have not split, otherwise water can build up and cause rusting. It is probably worth mentioning that the sunshine roof isn't prone to problems which is unusual for such devices. Moving to the front of the car, the bottom of the bulkhead should be examined for signs of rust and any bubbling paint may indicate that it is imminent.

The Y type's chassis doesn't usually rust though the point at which it passes directly beneath the rear axle is suspect and furthermore access to it is difficult.

Suspension, steering and brakes

The front suspension isn't troublesome, though if it has not been regularly maintained by greasing every 3000 miles it will wear very badly. Any wear is most likely to show itself by the distance tubes in the upper and lower swivel pin links seizing on their bolts. This elongates the holes at the ends of the top suspension arms and lower wishbones, which is hardly a desirable state of affairs. Although the links themselves and the swivel pins also wear, replacement parts are now available. The shock absorbers that form an integral part of the suspension do weaken and are awkward to replace.

The rack-and-pinion steering gear gives remarkably little trouble, though a gaiter check is necessary to ensure that they haven't split or fractured.

The brakes on the YA were shared with the contemporary Morris Ten and, though repair kits for the hydraulics are available, replacement wheel and master cylinders are rather more difficult to obtain. By contrast, the YB was fitted with the TD's twin leading-shoe brakes, which ensures a better supply of parts.

Engine, gearbox and transmission

The XPAG engine used in the Y series is a remarkably viceless unit but it is unfortunately rather difficult to keep oil-tight! It oozes out of the rocker-box cover and engine side-plate but consumption isn't too heavy—an average of about 500 miles per pint. An acceptable oil pressure reading for this engine is anything between 40 and 70psi at 40mph. A minor modification was made from engine number XPAG/SC 15405 when the Purolator throwaway oil filter type PTX 346 replaced the original Wilmot Breedon one. However, the appearance of the original unit can be preserved with a throwaway interior as NTG Motor Services of Ipswich, Suffolk market a very practical conversion. Surprisingly, these engines have metric threads, though the nuts have Whitworth flats—a reminder that Morris Engines in Coventry started as a branch of the French Hotchkiss company and the old metric dies continued to be used right up until 1955.

The gearbox is the same as that used in the TD and TF 2–seaters and is a rather fragile component. Some of the cogs can be noisy, particularly second gear.

Moving down the drive line we come to the rear axle. Unfortunately, the YA employed the Morris Ten Series M unit and this does have a habit of consuming half-shafts. Differentials are difficult to obtain and if it is possible to drive the vehicle under inspection then one should be alert for suspicious noises from the rear. Fortunately, the YB, which used the TD's hypoid unit, should present no major problems.

Spares

The Y type fraternity is fortunate that many mechanical parts for the model are available

***Far left*, the YA's instrument panel with the inevitable octagonal dials. *Above, left*, spare wheel access, nowadays a bad rust point. *Left*, the YA's XPAG engine. The hydraulic jacking unit can be seen on the right, below the radio vibrator unit**

What better two seater transport for an evening out in 1955 than an MGA? Unlike the later coupé, the roadster had no external door handles. *Right*, the Mark II version of the MGA 1600 had this insert radiator grille

because they are shared with the TD and to a lesser extent the TC. Therefore, engine, gearbox and suspension spares are relatively easy to come by. NTG Motor Services of Ipswich have taken the initiative and had a number of those rather tricky rubber trim parts re-manufactured and can also supply a full range of mechanical spares. Other companies which supply parts are Toulmin Motors of Brentford, and Moto-Build of Hounslow, Middlesex. Clubs are the MG Car Club, which has a Y type register, and the MG Owners' Club.

Spares and services

Moto-Build Ltd., 128 High Street, Hounslow, Middlesex (tel: 01-570 5342).
NTG Motor Services Ltd., 21, St Margaret's Green, Ipswich, Suffolk (tel: Ipswich 211240).
Toulmin Motors (1962) Ltd., 103–105 Windmill Road, Brentford, Middlesex.

MG Car Club Ltd., 67 Wide Bargate, Boston, Lincolnshire PE 21 6LE (tel: 0205 64301); MG Owners' Club, 13 Church End, Over, Cambridgeshire (tel: Swavesey 31125/30661).

Performance and consumption
MG Y type, *The Motor*, May 28, 1947.
Top speed 69mph. Overall mpg: 27 for 1020 miles.

MG Y type production

YA saloon	6158
YB tourer	877
YB saloon	1301

MGA 1500 1955–59 1600 1959–62

Specifications and developmental history

The MGA Roadster went into production in September 1955. It had an open, 2–door, 2–seater body mounted on a box-section chassis fitted with coil-and-wishbone independent front suspension. Lockheed hydraulic drum brakes were used and Cam Gears rack-and-pinion steering was employed. The engine was the BMC B series — a pushrod, overhead-valve, 4–cylinder of 1489cc (73 × 89mm) developing 68bhp. The bucket seats were leather upholstered and the car sold for £844.

In September 1959 a new hood with quarter lights was introduced (chassis 20162). A coupé version appeared at the same time with a larger wrap-around windscreen and wind-up windows. Both models were discontinued in May 1959.

The MGA 1600 was introduced in May 1959 with a 1588cc (75 × 89mm) version of the existing engine developing 80bhp. Lockheed disc brakes were fitted to the front wheels. New flashers with separate housings, instead of a combined unit, were introduced. Roadster and Coupé ceased production in March 1961.

A Mark II version of the 1600 appeared in June

1961. The engine's capacity was increased to 1622cc (76 × 88mm) developing 93bhp. A redesigned insert radiator grille was a distinctive feature. Production terminated in June (Coupé) and September (Roadster) 1962.

Check points

Body and chassis

The MGA has a separate frame but the condition of the bodywork is, nevertheless, important. Although steel predominates, the door, bonnet and boot lid are aluminium, so searching for rust in those areas is unnecessary! The examination should commence with the front wings. Their lower edge is a common rust trap, while the nearby door pillars are another vulnerable area. The condition of the doors can be assessed by opening them and trying to move them up and down — they shouldn't move. If there is any perceptible movement then it should be ensured that the hinges are not at fault. If it is then the pillars have probably seen better days and they are expensive items to replace. The front of the rear wheel arches should be examined for rust, filler, or both, while the extreme rear of the boot floor is another area that can be easily overlooked. The coupé version suffers where the fixed head joins the body, the cracking being caused by body flexing. Condition of the radiator grille is important since it is expensive to replace, particularly so on the 1600cc cars.

The chassis, a sturdy box-section frame, can be rusty. A particularly vulnerable area is the inside edge of the main chassis members immediately adjacent to the wooden floor-boards. These are difficult points to check because they are accessible only from inside the car and the suspect areas are usually covered by carpet. The plywood floor may be in very poor condition, but this is hardly a structural disaster. The battery carriers, which are slung either side of the prop-shaft behind the front seats, often suffer from corrosion. The parts of the chassis where the tubular cross members join the main frame should be examined for excessive rusting. Right at the back of the car the metal straps securing the petrol tank may have

rusted through and snapped.

Suspension, steering and brakes

The A's weak point is, undoubtedly, its independent front suspension, which requires very regular lubrication if the vehicle in question is in regular use; every 2 or 3 months is essential if this is the case. However, most owners are not that conscientious so one should be alert for plenty of wear in the top and bottom trunnions. Fortunately, replacements are available. It is also worth noting that the lower link spacer tube and washers are the same as those used on the MGB.

The steering gear is fairly trouble-free but the rack's gaiters should be examined for signs of damage. The unit is mounted right at the front of the car so it is subjected to a great deal of mud and stone bombardment.

There are no particular problems with the all-drum braked cars, but those fitted with front discs do have some unique snags. The chromium-plated calipers are particularly prone to seizure even if the vehicle in question has only been off the road for a comparatively short time. Jacking up the appropriate wheel will reveal whether it is binding, while a scored disc is another indication of trouble of this nature. The rear suspension is fairly trouble-free (the shock absorbers certainly last longer than the front ones) but the leaves are prone to breakage, so they should be inspected carefully.

Left, **cockpit of an MGA 1500. The central dashmounted horn button was an unusual feature.** *Above*, **under bonnet view of a left hand drive 1500**

Engine, gearbox and transmission

The faithful BMC B series engine, does not suffer from any particular problems. However, a reasonable oil pressure reading is important and this should not be less than 50psi at 50mph (hot). Fuel economy is a big advantage with this car, but it does, however, depend to a great degree on the condition and synchronization of the twin H4 SU carburettors.

Similarly, the A's gearbox is generally a robust and reliable unit, though it is possible that the synchromesh on second gear has failed. The rear axle is also reliable.

Spares

MGA owners are able to obtain both replica body panels and mechanical components. Body panels are obtainable from Moto-Build, while this company also supplies mechanical spares along with Toulmin and NTG Services, details of which can be found on page 189. In the north there is Simon Robinson's MGA Centre at Darlington and in London there are A specialists Vic Ellis Sportscar repairs and Roy McCarthy's Custom Style Autos, which caters for mechanical and body restorations. Clubs are MG Car and MG Owners', already mentioned on page 189.

MGA spares and services

Vic Ellis Sportscar Repairs, 234 Trussley Road, Hammersmith, London W6 (tel: 01-741 2731).
Custom Style Autos, 1a Windermere Avenue, London (tel: 01-540 7028).
Simon Robinson, Cleveland Street, Darlington, Co. Durham.

MGA performance and consumption (1500), *The Autocar*, September 23, 1955.
Top speed: 98mph. Overall mpg: 27 for 672 miles.

MGA production

1500	58,750
1600 Mark I	31,501
1600 Mark II	8719

MGA Twin Cam 1958-1960

Specifications and developmental history

The MGA Twin-Cam model was introduced in July 1958 in Roadster and Coupé form. Dunlop disc brakes were fitted all round with disc centre lock wheels from the same manufacturer. The BMC B series-based, twin-overhead camshaft engine was of 1588cc (75 × 89mm), developing 108bhp. The Roadster was priced at £1266. A number of modifications were made to the Twin Cams during their 2–year production life. Chassis changes were: wheel arches fitted with detachable panels (592); water temperature gauge changed from Fahrenheit to centigrade reading (713); flasher changed from white to yellow at front and separate unit introduced at rear; also minor trim changes in Coupé (2192); anti-roll bar fitted (2275); half-shaft and differential fitted with finer splines. These also apply to the De Luxe model (2371). Engine changes were: dynamo pulley changed to smaller size requiring shorter fan belt (engine number 272); gearing on oil pump changed from nine to ten teeth on half speed shaft; piston rings changed from chrome plate to iron (446); packing plate fitted between engine mounting and chassis (528); tappet bucket increased in length from 1.25 to 1.5 inches; heavier conrods fitted (1343); tappets sleeved into cylinder head (1587); distributor changed to non-vacuum advance type (2223); valve springs changed (2251). Production ceased in the spring of 1960.

Late in 1959 the MGA 1600 De Luxe appeared. This was a hybrid model, utilizing the Twin Cam chassis, but fitted with the 1588cc pushrod engine. Very few were produced and the model was discontinued in 1960. The Mark II 1622cc cars were offered unofficially with De Luxe specifications in June 1961 but nobody is very sure when the option ceased and how many were made!

Below, **there was little to distinguish the 1958 Twin Cam from its pushrod sister apart from the handsome Dunlop wheels and badges above the bonnet side air intakes. But it was £270 more expensive at £1265.** *Right*, **the 1588cc BMC B series based engine. Note the distributor on the opposite side to the pushrod cars, header tank and long dipstick**

Check points

Body and chassis

The same remarks apply to the Twin Cam cars as those already detailed.

Suspension, steering and brakes

Similar checks to those described for the pushrod cars should be conducted on the suspension and steering. The Twin Cams differ in that they are fitted with Dunlop disc wheels, which are located on four pegs, rather than splines as might be expected. Dunlop disc brakes are fitted all round and they do have their own problems; the rear discs tend to rust badly and the handbrake doesn't function particularly well! Difficulty in obtaining spares for the Dunlop master cylinder may also be experienced. Rubbers are no longer available and the only solution is to fit a non-standard unit.

Engine, gearbox and transmission

The Twin Cam's engine is rather more problematic than its pushrod sister. When the model first went on sale in July 1958, the first 345 examples were fitted with pistons carrying chromium-plated rings. Excessive bore wear resulted and the plating was dispensed with. It is unlikely that any of these original engines are still in circulation, but it does illustrate the nature of the difficulties when the model was new. An important modification came at engine number 1587 when the tappet buckets were steel-sleeved into the aluminium cylinder head. Prior to that they ran without sleeves and excessive wear occurred. Also it is worth noting that new camshaft bearings are virtually impossible to obtain!

Originally the Twin Cam ran at a 9.9:1 compression ratio though this was later reduced to 8.3:1. If a Gold Seal replacement engine is contemplated then it will almost certainly employ this lower compression. Also the engine functions much more efficiently with the vacuum advance disconnected and a non-vacuum advance distributor was one of the last modifications made to this engine, as excessive advance contributed to piston burn-out. An acceptable oil pressure reading is 50psi at 2500rpm, which is about 35mph in top gear. Not surprisingly an oil consumption figure of about 200 miles to the pint is about average.

Other minor difficulties with this engine occur when a fan belt change becomes necessary. This is because the radiator has to be removed to permit access to the lower adjusting nut on the dynamo, though removable panels were fitted inside the wheel arches of later cars. Should the car's cooling system require draining the block must be attended to separately. And unlike the pushrod version the Twin Cam is fitted with a separate header tank.

A last word on this engine. A petrol consumption figure of 20–25mpg is normal and it isn't worth improving on as weakening the mixture can result in burnt out pistons. The engine is noticeably efficient on Champion N4 sparking plugs, incidentally.

The gearbox shouldn't present any major problems but there could be difficulties with the rear axle on a Twin Cam later than chassis number 2371. The unit was fitted with a differential and associated half-shafts with a greater number of splines than employed hitherto. If the vehicle under inspection has troubles of this nature they are difficult to overcome because the parts were not used on any other MG model or other BMC car for that matter.

Spares

Many of the Twin Cam's parts are common to the pushrod models, the main exceptions being engine, gearbox, transmission and wheels and brakes. Peter Wood of Twyford, Bucks has a comprehensive selection of Twin Cam spares and also keeps pushrod parts as well. Clubs are mentioned on page 189.

MGA Twin Cam spares and service

Peter Wood, Westwood, Church Street, Twyford, Buckinghamshire (tel: 029673-310).

Twin Cam performance and consumption, *The Motor*, July 16, 1958.
Top speed: 113mph. Overall mpg: 22.2 for 1593 miles.

Twin Cam production	2111

Morgan 1910 to present date

Henry Frederick Stanley (H.F.S.) Morgan (1881–1959) was a clergyman's son who, after education at Marlborough and the Crystal Palace Engineering College, joined the Great Western Railway in 1899 as a pupil under Locomotive and Carriage Superintendent William Dean. But he decided against a permanent career in railways and left the GWR's Swindon works in 1906 moving west to Malvern Link where he opened a garage in Worcester Road selling Darracq and Wolseley cars.

Morgan soon entertained ideas of building a vehicle of his own and acquired a 7HP Peugeot V twin engine, which he first contemplated using in a motorcycle. The vehicle that finally emerged, however, was a 3–wheeler built with the help of Stephenson-Peach, engineering master at nearby Malvern College. Work began in 1908 and what was to be the prototype Morgan was completed early in 1909. It was an ingenious, yet simple, design of tubular and, therefore, light construction. The ladder chassis was built up around a central backbone tube that also carried the propeller shaft, while the side ones also saw service as exhaust pipes. But the Morgan's most important feature was the sliding pillar independent front suspension, a progressive design for the time. Such was the soundness of this original concept that this suspension system has been used on every subsequent Morgan, a tribute to HFS's advanced thinking. The engine was the aforementioned Peugeot positioned between the front wheels, the shaft drive being transferred to chains just ahead of the single rear wheel.

Although at this stage Morgan had little interest in marketing his car, it aroused such keen local enthusiasm that he exhibited two examples at the first Cycle and Motor Cycle exhibition at London's Olympia in 1910. Response was so encouraging that he decided to put the car into production. His father provided £3000 so that he could extend his garage premises and the Morgan Motor Company was founded in 1912. Demand soon grew, but the outbreak of the First World War in 1914 saw the start of a gradual switch to munitions and afterwards, in 1919, production was transferred to a new factory at Pickersleigh Road, about half a mile from the original works.

The cars, mostly powered by JAP, Blackburne and later Matchless engines were produced, though the F model of 1934 saw the introduction of a Z-section chassis frame, made by Rubery Owen, but which was not used on the V twins. Power for this new model came from a 4–cylinder Ford 8 engine with a 10HP De Luxe option for 1936. The same year saw the appearance of the first Morgan 4–wheeler, being called the 4/4. This indicated 4 wheels ınd a 4–cylinder engine, in this instance a 1122cc Coventry Climax unit. Production continued until the outbreak of the Second World War, though in 1939 a special 1276cc overhead valve version of the Standard Flying Ten engine replaced the Coventry Climax products.

During the Second World War the company manufactured parts for the Oerlikon gun together with various aircraft undercarriage components, gauges and compressor parts. Standard engined 4/4 and Ford 8 and 10HP–powered 3–wheeler production commenced after the war, but the export drive virtually spelt the end of the ageing smaller model, which was finally phased out early in 1952. A further complication came when Standards announced that they were only going to build one postwar model: the 2–litre Vanguard saloon. Thus, the 4/4 had to be modified to accept the larger engine and the outcome was the Plus 4 of 1951, which used the 2088cc Vanguard power unit. For 1954 the model was fitted with the 1991cc TR2 engine, the Plus 4 subsequently benefiting from updates in the TR sports car range.

The 4/4 was revived for 1956. This Series 2

model used the 1172cc side-valve Ford 100E engine. H. F. S. Morgan died in 1959 just missing the 50th anniversary of the announcement of his 3-wheeler, though from the late '30s the day-to-day running of the company had been in the capable hands of George Goodall with HFS's son Peter taking over as deputy governing director in 1951. The late '50s were difficult years for the Morgan company as their products appeared dated and generated little interest on the home market. In fact, in 1960, no less than 85 per cent of output was exported to America. The following year even that previously buoyant sector suffered a drop in demand and Malvern were forced to curtail their production for a couple of years. Valuable publicity was gained with a win at Le Mans in 1962 in the 2-litre class; the vehicle was the Super Sports Plus 4 that had been introduced for 1961 with lower body lines and Lawrencetune engine. Then, for 1964, came an uncharacteristic creation for Morgan — the Plus 4 Plus with a glass-fibre coupé body. The model was unsuccessful and was discreetly dropped after 4 years. Fortunately, from the mid-'60s demand improved again, particularly in the European market and since then the company hasn't looked back!

Meanwhile, the Ford engined 4/4 was soldiering on and the 1961 season saw the appearance of the Series III version powered by the 997cc 105E power unit. Then, that year's Motor Show witnessed the arrival of the Series IV for 1962, though this time the 109E 1340cc engine was fitted and front disc brakes were standardized. The Series V car of 1963 was upped in power by the fitting of the 116E 1498cc engine. Maximum speed was approaching that of the Plus 4 and the gap was further narrowed by the 4/4 1600 of 1968.

In 1966 Rover made a surprise take-over bid for Morgan, offering to supply their then secret Buick-derived V8. But Peter Morgan, who had become accustomed to approaches over the years, politely declined just as his father had back in 1950 when approached by Standard's Sir John Black. In both instances (as it transpired) approaches came from engine suppliers but Morgan's response made no difference to supply. Rover, in fact, loaned a non-working V8 for Morgan's design purposes. Then came a working Buick unit.

In 1968 Peter Morgan signed an agreement with General Motors allowing his company to use the 3500cc Rover V8 which, it will be recalled, was based on a Buick design. The new model, called the Plus 8 was launched at the 1968 Motor Show and this, along with the 4/4 forms the basis of the company's products to this day (1980).

Morgan production 1946–1967

V twin	12
F type (Ford 8 and 10 engined)	243
4/4 Series I	534
4/4 Series II	486
4/4 Series III	60
4/4 Series IV	218
4/4 Series V	625

Morgan Plus Four: top speed 102mph, 1954 price £829

Plus 4 1951–1969

Specifications and developmental history

The Morgan Plus Four was announced in open two–seater and drophead coupé form at the 1950 Motor Show. It used an underslung Z-frame chassis with sliding-pillar independent front suspension. Girling hydraulic drum brakes were fitted all round. Cam-and-peg steering was employed. The engine was a Standard Vanguard unit of 2088cc (85 × 92mm) developing 68bhp. The seats were upholstered in leather. The cost of the open two–seater was £652 while the coupé was £722.

A four-seater tourer joined the range for 1952. The two–seater's front end was redesigned and a partly cowled radiator and headlamps faired into the front wings were announced at the 1953 Motor Show. In March 1954 the coupé and 4–seater tourer followed suit. At the same time the 90bhp 1991cc TR2 engine became available at £20 extra and was standardized on the two–seater for 1955. Meanwhile the Vanguard engine was available until 1957 and the following year all models were fitted with the 100bhp TR3 engine.

Cars for 1956 incorporated a few chassis changes: the front brake drums were wider; two of the greasing points were eliminated and the sliding rear-spring trunnions were replaced by rubber bushes. Cars for 1958 used an improved radiator cowl 3 inches deeper, the two-seater's seats were widened 4 inches and the running boards narrowed accordingly. The 1959 Plus 4's tail was altered to bring it into line with the Series Two 4/4. The spare wheel was, therefore, recessed into a sloping rear panel rather than carried vertically. The fuel tank was moved lower down the chassis and the capacity increased from 11 to 12 gallons. The 4–seater's bodywork was also widened, bringing it in line with changes to the two–seater made the previous year. Mechanical developments included the introduction of Thompson tubular rods and joints for the drag links and tie rods, while a heavier gauge metal was specified for the lower of the two front suspension tubes. In May 1959 Girling disc brakes were introduced with 15–inch wheels appearing as an optional extra and eventually standardized for the 1960 season. In 1961 the model's rear axle was changed from a Salisbury 3HA unit to a 7HA one.

March 1961 witnessed the introduction of the Plus 4 Super Sports with Lawrencetune 115bhp engine and dual-choke twin Weber carburettors, four-branch exhaust manifold and wire wheels. In

Below, **the rarer Plus Four coupé of 1951, with earlier flat fronted radiator**

The 1954 Plus Four showing partly cowled radiator and faired in headlamps

July 1962 the TR4 engine of 2138cc (86 × 92mm), developing 105bhp, was introduced to the range. From January 1963 the coupé model was available on special request. The Plus 4 Plus appeared at that year's Motor Show, with glass-fibre coupé bodywork. The TR4A 105bhp engine appeared in 1965. The range included the Plus 4 Competition Model with 4/4 body styling and 112bhp engine, though this and the Plus Four Plus were dropped at the end of 1966. For the 1967 season the Plus 4's bodylines were lowered, like the rest of the Morgan range. Production ceased in January 1969.

Check points

Body and chassis

Practically all the Plus 4s have steel bodies, except the Super Sports which has an aluminium one. At the front of the car, the wings around the side-light mountings should be examined as vibration from the sliding pillar suspension beneath often causes cracking. Next for inspection is the bodywork below the doors and immediately adjoining the running boards. Whilst open, doors should be checked for upward or downward movement. Early examples were fitted with brass hinges and steel pins so some wear can be expected.

This mid-point region does suffer from rusting underneath with the Z frame chassis corroding on the *inside*. Unfortunately, the seat must be removed to inspect it properly. The problem here is that water leaks under the door, is soaked up by the horse hair under the trim, which subsequently triggers rot in the ash frame and so on. Whilst probing the chassis, the points where the three cross-members join the main chassis frame should be checked as it is a common rust spot. Engine mountings should also be looked at as the car tends to vibrate somewhat.

Above, rear view of the 1954 car. *Right*, the 'office' of the same car though a flat steering wheel in place of the dished one was normal wear for the year

Right at the back of the car, the half round metal beading that runs down along each side should be examined. Finally, facing the side of the car, the rear wing should be grasped and lifted gently; the body should be firmly secured to the chassis, but if it isn't then the coachwork will come away from the frame. If this occurs then the vehicle should be avoided.

Finally, a word on the paint used on the model. Some early examples used sprayed coachpaint which was later replaced by cellulose. If a section of body is resprayed then one should be certain that the right paint is being employed as the two types interact. If there is the slightest doubt, a rag dipped in cellulose thinners can be applied to an unobtrusive area of bodywork. If the paint comes off then, the car in question has been sprayed in cellulose and this is what must be used subsequently. Alternatively, if the cloth is unmarked then the car was finished in coachpaint and some will be needed for touching in or respraying individual areas.

Suspension, steering and brakes

The Plus 4's most vulnerable area for wear is the independent front suspension system. Ideally it is oiled by a foot-operated, one-shot lubricator every 200 miles (more than this can be harmful to the brakes). In view of this rather haphazard method of lubrication, the kingpins and bushes will probably need replacing every 15,000–18,000 miles. However, they are modestly priced. The condition of the suspension should be checked in the usual way, but it should also be remembered that on this model the steering damper mountings also wear. These dampers (consisting of a flat steel spring, fixed to the chassis at one end and to the suspension via a phosphor-bronze bush at the other) prevent any twisting motion caused by the springs when under compression, which can be transmitted to the stub axles and so upset the steering. The oil level in the steering box should be checked every 5000 miles.

At the rear of the car the nuts on the U-bolts securing the springs to the rear axle should be checked periodically for tightness.

The braking system presents no peculiar problems.

Engine, gearbox and transmission

The Plus 4 is powered by the 4–cylinder engine used in the TR range of sports cars and check points for this particular unit can be found on page 255. However, it is worth noting that petrol consumption was never as good in the Morgan as when the engine was fitted in its own chassis.

A Moss gearbox is employed, being separated from the engine by a long electron transmission housing. This contains the drive shaft (with an aluminium sleeve between it and the housing) operating the clutch through a graphite thrust bearing. It is important that this bearing be lubricated every 3000 miles and if the clutch is noticeably heavy to depress whilst driving, it is very likely that the lubrication has not been regularly attended to.

The Salisbury rear axle has no particular disadvantages, but parts for the earlier 3HA unit are now becoming difficult to find.

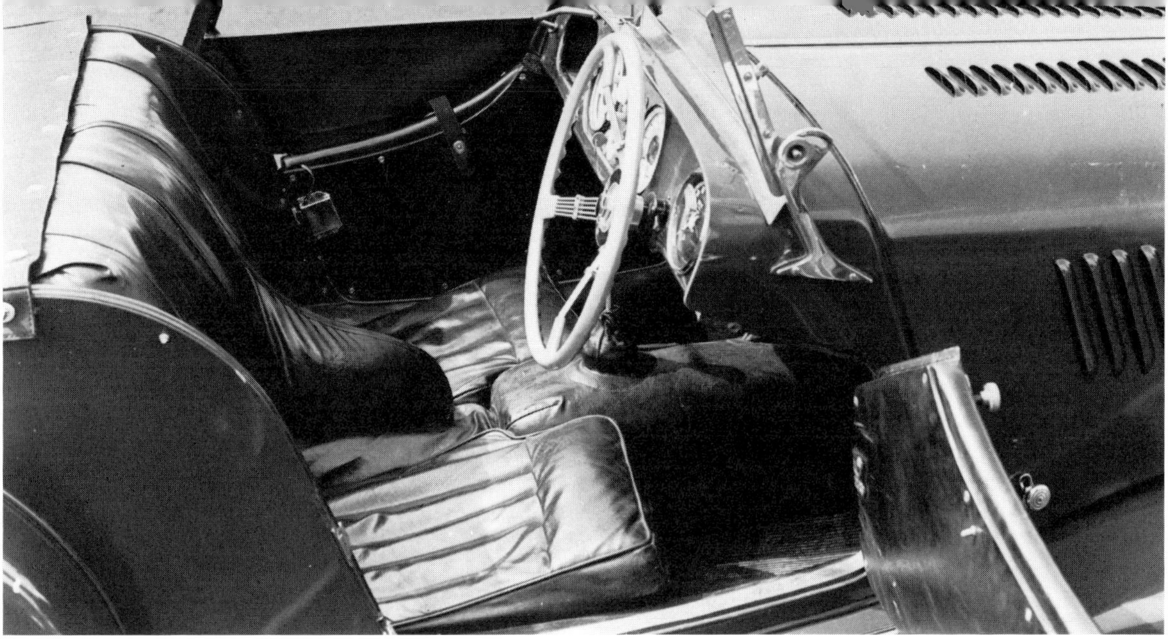

Spares

The fact that the Morgan factory is very much alive and well means that most spares, both body and mechanical are available for these cars. Other parts sources include Burlen Services of Salisbury, Wilts, F. H. Douglass and Melvyn Rutter Specialist Services of London. TR engine parts sources can be found on page 255. Another source is the ever-active Morgan Sports Car Club!

Plus 4 parts and services

Burlen Services, Greencroft Garage, Greencroft Street, Salisbury, Wiltshire (tel: Salisbury 21777).
F. H. Douglass, 1a South Ealing Road, Ealing, London W5 4QT (tel: 01-567 0570).
Morgan Motor Co. Ltd., Pickersleigh Road, Malvern Link, Worcestershire (tel: Malvern 3104/5).
Melvyn Rutter Specialist Services, 3 The Green, Wanstead, London E11 2NT (01-989 1287).

Morgan Sports Car Club, Charles J. Smith, 23 Seymour Avenue, Worcester (tel: 0905 52995).

The Standard Vanguard engine of a 1951 Plus Four, developing 68bhp at 4200rpm. An air filter was not fitted

Morgan Plus 4 production

2–seaters	348
4–seaters	141
2–seater coupés	117
2 + 2 coupé	1
TOTAL	607
Chassis only	51

Morris 1913 to present date

Top models! A 1937 four door Morris Eight, the best selling car of the 1930s. This Series I car is identified by its wire wheels and chrome radiator surround, all for £142. Series II cars had a painted radiator and Easiclene wheels

William Morris, Britain's most successful motor manufacturer of the inter-war years, sold his first car — the 2–seater Oxford — in 1913. It was joined in 1915 by the 1496cc Cowley with a power unit by Continental Engines of Detroit. Production continued in 1919 with Hotchkiss at Coventry manufacturing 1548cc copies of the American engine and this powered the best-selling British car of the 1920s. The 'Bullnose' Cowley and better equipped 1802cc Oxford remained in production until 1926 when they were replaced by the less inspiring 'Flatnose', which was available until 1934, though the Oxford option was dropped in 1929. The company's first successful six appeared for 1928 with a 2.5–litre, overhead-camshaft Coventry designed engine. For 1930 the six changed to a side-valve, taking the Oxford name, and the overhead-camshaft engine was transferred to the Isis. Both models were dropped in 1935. Morris entered the small-horsepower market in 1929 with the 847cc Minor, also carrying an overhead-camshaft power unit, though this was supplemented and in 1931 later succeeded by a more conventional side-valver; production lasted until 1934. A 10HP 4–cylinder car appeared in 1933 but this, like many other models was discontinued in 1935 with Len Lord, Morris's new managing director, bringing a sense of order to the unwieldy range.

In 1935 the marque scored a huge success again with the Eight, employing a 918cc side-valve engine. In mid-1935 the company abandoned their traditional model changes at the Motor Show instigating the Series versions, designed to give the cars a facelift in the traditionally slack pre-Autumn months. The Series II Eight appeared in August 1937, ceasing in 1938, 221,000 had been produced, making it the best-selling model of the decade. Ten and Twelve models, were announced in August 1937, and featured overhead-valve engines, though from this date the twelve emerged with its own identity. In August 1938 came the Series M Ten, a landmark for the company as it represented Morris's first foray into unitary construction bodywork. In October the same year a new version of the Eight, the Series E, appeared with updated styling and headlamps faired into the front wings. The engine was based on the earlier 918cc unit but with redesigned cylinder head, counterbalanced crankshaft and shell bearings. During the Second World War the Nuffield Organization produced a tremendous variety of items, such as Tiger Moth aircraft and Crusader tanks, and mine sinkers which had also been produced in the 1914–1918 conflict. At the

What district nurses used before there were Morris Minors; a Series E Morris Eight of 1947

end of the War Morris introduced just two models, his pre-war Series E and Series M Ten, though these were dropped in 1948 when a new model range was announced at that year's Motor Show.

These were the Minor, Oxford and Six, which all shared similar body styling, designed by Alec Issigonis and A. V. Oak. All were unitary or monocoque construction with torsion bar-and-wishbone independent front suspension and rack-and-pinion steering, except the Six, which used a Bishop cam unit. Power units ranged from the straightforward to the complicated. The Minor, designated MM, used the pre-war Series E Eight engine, though 800cc and 1100cc flat fours had been tried, while Wolseley were anxious to fit a 750cc version of their overhead cam engine, a similar package to the original Minor of 1929. A side-valve engine, though, of 1476cc also featured in the Series MO Oxford but the Series MS Six used a 2.2-litre, overhead-camshaft, 6-cylinder power unit. With the exception of the bonnet and traditional radiator uncomfortably grafted on to the fashionable bodywork, the Six shared styling and mechanics with the Six-Eighty Wolseley, announced simultaneously. Both these models and the Oxford were discontinued in 1954, though the Minor was produced right up until 1971. In 1952 it became the Series II, receiving the 803cc overhead-valve engine from the A30, following Morris's merger with Austin that year. By 1961 1,000,000 had been sold—the first British car to achieve this—and it is a reminder of those export-conscious years that no less than 48 per cent were sold overseas. By the time the last car appeared, late in 1970, over 1,500,000 had been built.

Although the design of the Series II Oxford of 1954 originated from Cowley, the engine, an overhead-valve, 1489cc, 4-cylinder B series, was again an Austin unit, though a smaller capacity 1200cc Cowley was also offered in the time-honoured Morris traditions. However, it received the Oxford's engine in 1956 and this remained unchanged until production ceased in 1959. The Series I Isis was introduced for 1956 with similar bodywork but fitted with the Morris designed 2.6-litre BMC C series engine. This went into Series II form, rather surprisingly, with right-hand gear-change for 1957 but was discontinued the following year. The Oxford, meantime, had gone to Series III in 1957, while the Series IV of 1957–1960 was the Traveller version with opening tailgate. In 1959 came the Series V which was a Farina-styled saloon, shared with other BMC marques. The longer wheelbase, wider track, 1622cc Series VI appeared for 1962, the model finally being dropped in 1971, though it was also diesel engined between 1964 and 1970.

The Morris version of the revolutionary front-wheel-drive Mini Minor was announced simultaneously with the Austin one in 1959, though the Farina-styled 1100 with Hydrolastic suspension of 1962 initially carried the Morris name plate. In 1966 came the Morris equivalent of the Austin 1800 and for 1968 the 1100's capacity was raised to 1275cc and designated the 1300.

Morris production 1945–1958

Eight Series E	67,534
Ten Series M	53,570
Six Series MS	12,465
Oxford Series MO	159,960
Oxford Series II	87,341
Oxford Series III and IV Estate	57,117
Isis	6692
Isis Series II	6414
Mini Cooper (to 1971)	100,051
Mini Cooper S (to 1971)	45,442

Minor 1949-1970

Specifications and developmental history

The Morris Minor was introduced at the 1948 Motor show in 2–door saloon and tourer form. Monocoque construction was employed. Torsion bar-and-wishbone, independent front suspension was used. Lockheed hydraulic brakes were fitted, while the steering was a rack-and-pinion unit. The engine was a 4–cylinder, side-valve 918cc (57 × 90mm) from the Morris Eight Series E and developed 27bhp. Seats were covered with Vynide, though leather was also available. Both models sold for £358.

In June 1949 twin tail-lamps were introduced. With the appearance of the 4–door saloon for 1951, the front headlamps were moved from the radiator grille to the front wings, separate sidelamps being introduced at the same time. Those cars fitted with a heater used a water pump (chassis number 63822). The headlamps were also moved on the 2–door saloon and tourer for January 1951 (83390). In March of that year a painted radiator grille was introduced in place of the plated one (90177) and in June the tourer received fixed quarter-lights instead of the detachable side-screens used previously (1000920).

In July 1952 the Series II Minor was introduced and the 4–door saloon (16001) was fitted with the

Minor variations: *Below*, **1948 Series M M saloon with grille mounted headlamps and divided windscreen.** *Bottom left*, **1960 Minor 1000 with 948cc engine. The split screen disappeared in 1957.** *Right*, **the tourer version, note the same registration number!**

803cc (58 × 76mm), overhead-valve, pushrod engine (developing 30bhp) from the Austin A30. It also appeared in the 2–door saloon and tourer in February 1953 (180001). For 1954 a Traveller version was added to the range with a composite wood and metal body, the new three-quarter floating rear axle being fitted to all models. A new grille with horizontal bars was introduced for all models for the 1955 season and a single mounted instrument fitted in place of the separate speedometer, fuel and oil pressure gauges used previously (289441).

The 1957 models were designated the Morris 1000 having a more powerful 948cc (62 × 76mm) A series engine, developing 37bhp. Other changes included a one-piece windscreen in place of the distinctive divided one, a larger rear window, remote control gear lever, lids fitted to the glove boxes, dished type steering wheel with the direction indicator switch moved from the facia to the steering column (448801). In February 1957 the canvas hood on the tourer was replaced by plastic material (477961). In January 1961 the one millionth Morris Minor was built and replica examples were adorned with *Morris 1000000* in place of the *Morris 1000* legend. Cars for the 1962 season were fitted with flashing direction indicators instead of the semaphore units used previously, while screen washers were fitted on the De Luxe models.

Vehicles for 1963 were fitted with a larger capacity A series engine of 1098cc (64 × 83mm), developing 48bhp. Clutch diameter was increased to 7.25 inches, while the front brake drums were enlarged to 8 inches diameter. A fresh-air heater appeared in place of the recirculating type in April 1963. In June new amber flashing indicators were introduced. Sealed-beam headlamps were included for the 1967 season. In September of that year the Tourer was discontinued. December 1970 saw the saloon and Traveller cease production.

Check points

Bodywork

The first priority will probably be the front wings and rust may be detected breaking through around the indicator light, where this is fitted, and also around the headlamp bowl. The area of wing immediately adjoining the front door is particularly prone to rust on the Minor, but seems worse on the later cars. The bonnet is then

The Traveller was introduced in 1954. This is a 1960 Minor 1000 which sold for £669 though a heater was reserved for the de Luxe model

374 KFC

Driving compartment of an MM saloon with layout designed by Alec Issigonis

opened and the valances at the bottom of the engine compartment should be examined for rust. The top of the engine compartment can also suffer. With the front door open the outer sill is checked; its general state isn't too critical but a careful inspection of the inner sill under the car is important. Check also the condition of the cross-member; it has a vital function as the torsion bars are fixed to it and it also serves as a jacking point. If it is badly rusted then the floor directly above it may have deteriorated, corroding in a line right across the car, but it is usually worse at its extremities. Inside the car, the carpets should be raised and the edges of the floor pan checked for rust. On the two-door cars, rust can also take hold at the bottom of the quarter panel but this is more obvious on the outside of the car. The bottom of the doors suffer and on the two-door examples they tend to crack under the quarter-lights.

But there are even worse deficiencies! The rear wings alongside the doors rust, in much the same way as the front. The boot is also an important area and the floor should be examined, particularly in front of the spare wheel and in the corners. The lid itself may also be affected.

Suspension, steering and brakes

The front suspension is very durable, though if the swivel fails in the top suspension upright the front wheel will tuck in at an angle inside the wheel arch. This is usually attributable to lack of maintenance. The only way of checking that such collapse might be imminent is to jack the front of the car up and check for signs of greasing around the suspect swivel. If there are no signs of maintenance and if there was a rather a lot of 'drop' as the car's weight was taken off the wheel, this may indicate trouble ahead. This is the system's only really major shortcoming, but like other independent suspension units it is expensive to overhaul.

Although the rear springs are reasonably reliable, particular care should be taken to check the condition of their hangers, both front and rear. They are prone to collapse, so prod their immediate surroundings with a screwdriver.

The rack-and-pinion steering gear shouldn't suffer from any play and the rubber gaiters should be checked for splits. Brakes are fairly trouble-free, but the master cylinder is situated, rather surprisingly, beneath the floor on the driver's side and access to it is somewhat difficult.

Engine, gearbox and transmission

As already mentioned, the Series E side-valve Morris engine was fitted to the Minor up until 1953 and these have unique problems. Perhaps the main shortcoming is that a crack develops between the third cylinder head stud from the front to the nearest water hole. Often this can be overcome by refacing the head and block, but if the vehicle in question is suffering from top end problems, this is a likely cause. Also one should be alert for rusting of the water outlet pipe on the offside of the cylinder block but the problem can usually be resolved by fitting a slightly longer hose. Gearbox and rear axle are prone to some noise which could develop into something worse.

From 1953 onwards the A series engine was virtually the same as that fitted to the A30 and later the A35. Although parts are not usually a problem some difficulties may be experienced with spares for the earlier 803cc units. However, the power unit was phenomenally reliable and will continue to function even in an advanced state of exhaustion.

Spares

As might be expected, mechanical parts are not a great problem with the later Minors; the earlier models present most difficulties. Side-valve engine spares may be difficult to obtain while exterior trim is very scarce. It is certainly encouraging to know that new metal wings are still being produced, and these may well be available through the usual retail outlets. It is often worth checking with your local stockist. Restoration is undertaken

by the Morris Minor Centre of Bath. Fortunately, the Morris Minor Owner's Club monitors the supply of scarce parts and, in some instances, have initiated manufacture on their own account. So there is one good reason for joining.

Minor spares and services

Morris Minor Centre, Avon House, Lower Bristol Road, Bath, Avon (tel: Bath 315449).
Morris Minor Owners' Club, Paul Davies, 10 Beech Avenue, East Leake, Loughborough, Leicestershire (tel: East Leake 3226).

Minor performance and consumption (1000 4–door saloon), *The Motor* November 28 1956.
Top speed: 72.4mph. Overall mpg: 36.3 for 991 miles.

Morris Minor production

(Saloons, tourers and Travellers, including Abingdon built examples of the latter)

Series MM	159,500
Series II	288,714
1000 (948cc)	541,488
1000 (1098cc)	298,088
TOTAL	1,287,790

***Top*, the Minor's 918cc side valve engine, inherited from the Series E. *Above*, the 803cc engine from the A30 saloon, though with SU rather than Zenith carburettor, fitted to some Minors from 1952**

The balance of the oft quoted Minor production total of 1,582,302 is made up of light commercial derivatives, van and pick-up, $\frac{1}{4}$ ton/O-series/6–8 cwt, as they are variously known. These had totalled 150,000 from 1953 to 1962 and continued in production after the saloons were discontinued together with a badge engineered Austin variant.

Riley 1898–1969

Originally the manufacturer of weaving equipment, the decline in that industry forced the Riley family into the bicycle business and so into car manufacture. William Riley, Jr, who had taken over his father's Coventry firm, built tricycles, tricars and later, in 1906, 4–wheelers. Meanwhile, another sector of the company built up a lucrative business in detachable wire wheels which were sold to firms like Hispano–Suiza, Mercedes and Rolls–Royce. The company was re-shaped; bicycle manufacture was dropped and the car manufacturing side underpinned with the creation of Riley (Coventry) Ltd., in 1912.

After the First World War car production was re-started with a 1.5–litre side-valve model. In 1923, Victor, one of William's sons took over the managerial reins from his father. But the real milestone, in terms of the company's future prospects, came in 1926 with the appearance of the 9HP Monaco saloon. Sporting a high-waisted fabric body in the French manner, the Nine also featured a self-contained boot — a rare luxury in the 1920s. The power unit was even more noteworthy. Designed by another of William's sons, Percy, this four–cylinder, 1100cc engine featured a clever valve arrangement with two gear-driven camshafts set high in the block either side of the cylinder bores. Via short pushrods, these actuated overhead valves, inclined in a hemispherical cylinder head. The rocker boxes on top of the engine suggested twin overhead camshafts, but the power unit offered many advantages of this system, though without the cost or complication. Not surprisingly this ingenious engine was very popular with the tuning fraternity of the day and the company later pursued an ambitious racing and rallying programme. Such was the advanced design of this engine that the layout remained a feature of the Riley range right up until 1957.

A variety of models followed, a 6–cylinder Fourteen appearing for 1929 while the original 1100cc Nine was joined in 1934 by a larger 1.5–litre version, which was designed by Hugh Rose. This appeared, in the first instance, in the Lynx tourer and then in the Falcon and Kestrel saloons. In 1937 a larger 2.5–litre Big Four appeared though less successful were the products of a Riley subsidiary, Autovia Cars, who produced a few 2.8–litre V8s concocted from two 1.5–litre cylinder blocks in 1937–1938.

Many of these Rileys, along with the short-production Imp, Sprite and even scarcer MPH 2–seaters, possessed highly distinctive body styles in much the same way that a Jaguar is instantly recognizable today. Unfortunately, towards the end of the '30s individuality palled, profits plummeted and in February 1938 the company went into a voluntary receivership after considering a merger with Triumph, another Coventry company. Fortunately, the talks came to nothing (Triumph were bankrupted 12 months later) and in September 1938 Lord Nuffield personally purchased the firm for £143,000 following an appeal from Victor Riley, whom he had known since before the First World War. A month later the company became a subsidiary of Morris Motors being sold by Nuffield to his own conglomerate for a nominal sum of £100. Victor Riley remained as managing director and joined the Morris board. However, the models for 1939–1940 were largely Wolseley-based but Riley-powered which boded ill for the marque. During the Second World War the company became a shadow factory for the production of SU aircraft carburettors, while aero-engine mountings and Beaufighter undercarriages were among the other contracts undertaken. However, many of the dies for the pre-war cars were destroyed in the Coventry blitz, forcing the company to design a new model for the postwar era.

Consequently they were quick off the mark with their new $1\frac{1}{2}$ litre that employed the classic Riley engine, which was announced in 1945 in September — the same month as VJ day. Victor Riley's team, headed by Harry Rush, had produced a car worthy of the marque. It was a good-looking saloon, sporting a traditional radiator shell, while the fabric roof was distinctive. Its appearance was further enhanced when Sir Miles Thomas of the Nuffield Organization ordered a last minute reduction of 1.5 inches in roof height. The car's lines were based on a BMW saloon owned by a Riley employee before the war, which was involved in an accident and subsequently re-built with a Riley front end. Bert Holmes and Eric Carter were responsible for styling the new model, the coachwork for which was produced at the Morris Motors Coventry body works. It was constructed around a traditional ash frame, but then fitted with pressed steel panels with the bolt-on wings providing a vestige of past coachbuilding practice. The 4–cylinder, 1496cc engine was based on the 12/4 design of 1934, while the traditional Riley torque tube transmission was another concession to pre-war thinking. However, independent front suspension was a torsion bar-and-wishbone layout, while rack-and-pinion steering was fitted.

In November 1946 a 2½–litre version appeared. It was 7 inches longer than the smaller car and distinguished from it by a light blue radiator badge. (The 1½ litre car's badge was finished in a dark blue hue.) Like the smaller model, the engine also had pre-war roots, being basically the 2443cc power unit that first appeared in 1937. A Roadster version, aimed at the American market (with 20 gallon fuel tank), was announced in March 1948, but not many were made and the model was discreetly dropped in 1950.

Victor Riley was one of a group of directors sacked by Morris in the December purge of 1947 and attempts were again made to consolidate the unwieldy empire. Consequently, Wolseley production was transferred to Cowley, while Riley output moved to the MG factory at Abingdon, their Coventry factory being taken over by the Morris engine's division. Therefore, from 1949 Rileys were built alongside MGs with Riley man Jack Tatlow taking over as MG's general manager. Therefore, in February 1949, the company's name was changed from Riley (Coventry) Ltd., to Riley Motors Ltd.

A replacement for the 2½–litre, the Pathfinder (Series RMH), appeared for 1954. Designed by Gerald Palmer, it retained the Riley engine and torsion-bar independent front suspension and had a perimeter chassis frame, coil springs with trailing radius arms and Panhard Rod at the rear. A right-hand gear change was incorporated by Palmer to allow three people to use the bench-type front seat. Top speed was about 100mph, making the Pathfinder the fastest saloon in the BMC range. For 1955 the bodywork was shared with the Wolseley Six–Ninety and early in 1957 the rear coil springs were replaced by conventional semi-elliptics. The Pathfinder was finally dropped later in the year, marking the disappearance of the famous Riley engine, which had served the company for 31 years.

In the absence of any further individuality the marque slowly declined. For 1958 the Two-Point-

Drophead coupé RM Rileys, pictured on the occasion of the Society of Motor Manufacturers and Traders' Jubilee celebrations in July 1946. These may be the only 1½–litre versions ever made

Six appeared, with a basic Pathfinder body and chassis but carrying the BMC 2.6-litre, 6-cylinder C Series engine. (It is certainly worth noting that Palmer designed a twin-overhead-camshaft conversion of this engine, which was intended as a rival for the Jaguar. But only one example was built before the scheme was abandoned.) At the same time the Riley One-Point-Five was announced for 1958, the old 1½-litre model having been dropped in mid-1955. This used a Morris Minor floor pan and 1496cc B Series engine, and was a badge-engineered version of the newly introduced Wolseley 1500. The Four-Sixty-Eight of 1959 was a Farina-bodied saloon, which was shared by other BMC marques. In 1962 this car, with a 1489cc B Series engine, was increased in capacity to 1622cc to become the Four-Seventy-Two which remained available until 1969.

The last two models were revivals of pre-war Riley names, the Elf of 1962 being a luxurious version of the Mini, and featuring a larger boot. However, early in 1962 the engine capacity was increased from 848cc to 998cc, and the new version was designated the Mark II. Hydrolastic suspension appeared late in 1964, while wind up windows were a feature of the 1967 Mark III models. In 1965 the Kestrel name was revived as an upmarket version of the BMC 1100 model, becoming the Mark II with re-styled trim and rear lights; a 1300 version was announced at the same time.

In 1968 only 8346 cars had been sold and the famous Riley marque was the first to disappear in 1969 under the rationalization policy of the newly-formed British Leyland Motor Corporation.

Riley production 1954–1967

Pathfinder		5152
Two-Point-Six		2000
One-Point-Five Mark I		18,021
One-Point-Five Mark II		9777
One-Point-Five Mark III		12,084
	TOTAL	39,882
Four-Sixty-Eight		10,984
Elf (to 1970)		30,912

Above, **the Riley One-Point-Five of 1960, a variation of the Wolseley 1500.** *Right*, **the fine lines of a pre-1949 2½-litre RMB saloon**

$1\frac{1}{2}$-litre 1945–1955
$2\frac{1}{2}$-litre 1946–1953

Specifications and developmental history

The Riley $1\frac{1}{2}$–litre (Series RMA), initially called the Twelve, was introduced in September 1945. A box-section chassis was employed with torsion bar-and-wishbone independent front suspension. Girling Hydro-mechanical brakes (hydraulic front and mechanical rear) were fitted, while rack-and-pinion steering was specified. The engine was a 1496cc (69 × 100mm) unit with overhead valves inclined in a hemispherical cylinder head actuated via short pushrods from twin camshafts set high in the block. The engine developed 54bhp. Seats were upholstered in leather complemented by a wooden dashboard and door fillets. Price was £709.

In February 1946 the rear jacking points were moved from the running boards to the rear of the chassis. For 1949 internal bonnet locks replaced the external ones, the dip switch was moved from the floor to the dashboard and door pockets replaced the map pouches on the rear of the front seats. A rear seat armrest was fitted. The 1950 season saw the introduction of a new-style dashboard with imitation wood grain for the central instrument cluster. Gold-faced rectangular instruments replaced the pre-war circular white-faced ones. At the same time adjustable front seat armrests appeared (from chassis 17201). Other changes included the replacement of the all-steel steering wheel with a plastic-rimmed one, while the handbrake lever was moved from the left to the right of the steering column. Windscreen wipers operated under a single control, replacing the individual layout previously employed, the driver's opening window on the V-piece screen being deleted. Door fillets were plainer with inset bakelite ashtrays. In 1952 came redesigned bumpers, a one-piece full-width one at the front with double bar at the rear replaced the quarter bumpers previously used (19636). Improved seats were fitted and the ashtrays moved from the door fillets to the dashboard at the front and door panels

SHILLINGFORD BRIDGE HOTEL
EUE693

The three seater $2\frac{1}{2}$–litre RMC Roadster, specifically designed for the American market, pictured in MG country. But it failed to attract customers

at the rear. In April 1952 full hydraulic brakes were introduced while an open propeller shaft and hypoid rear axle were employed in place of the torque tube layout used previously. Changes for 1953 saw a higher roofline with larger rear window with chromium plated surround while the front seats were re-designed (again!) and the front seat armrests were deleted. A lower rear axle ratio completed the package and the model was re-designated RME (20505). For 1954 the running boards were deleted, spats fitted to the rear wheels and wings given knife-edge styling. Side and fog lamps were built in. Extra lamp clusters were added to the rear wings of the 1955 models, while in January of that year the D-style rear lamps were deleted and a centrally-mounted rear number plate was fitted. The model was discontinued in May/June 1955.

The $2\frac{1}{2}$–litre (RMB) was introduced in November 1946, being similar in style to the $1\frac{1}{2}$–litre, but with longer bonnet and wheelbase, the latter being 9ft 11in, compared with 9ft 4.5in. The engine was a 2443cc (80 × 120mm), 4–cylinder unit of similar layout to the smaller model and developing 90bhp. Price was £880. Modifications were similar to the $1\frac{1}{2}$–litre unless otherwise stated.

A Roadster version (RMC) with steering-column gearchange and three–seater layout was introduced for 1949 and this was modified as a two–seater with floor gear change in January 1950. It was discontinued in April of that year. A drophead coupé version (RMD) of the $2\frac{1}{2}$–litre also appeared for 1949 though this variant was dropped in October 1951. The changes of April 1952 saw the model being re-designated the RMB2, though the alterations for 1953 resulted in another re-designation to RMF. The $2\frac{1}{2}$–litre ceased production in October 1953.

Check points

Body and chassis

In the main, the RM's coachwork shouldn't present too many problems, though the exception is that distinctive and unusual fabric roof. Its most vulnerable area is where the rear of the roof joins the metal body just above the boot. The fabric should be pushed in with the thumbs. If it 'gives' this probably means that water has leaked in and rotted the wooden framework underneath. Also the parallel stitching along the length of the roof

RM saloons have well finished interiors with leather seats, wooden dashboards and door fillets. These map pouches were replaced by door pockets in 1949

should be examined. If this is in poor condition this is almost certain to be where the water has leaked in and unless a major body re-build can be undertaken then it would be advisable to avoid the example under inspection.

The metalwork also suffers from corrosion but this is usually cheaper and easier to rectify. The sector of the front wing immediately behind the headlights should be inspected very closely as this is a common rust trap, caused by mud and similar debris lodging underneath. Examine also the windscreen pillars; any signs of bubbling will indicate that all is not well below the surface and that the wood in that particular part of the car has seen better days. The rear wings should also be examined, particularly their forward ends immediately adjoining the rear doors. Whilst at the rear remember to inspect the floor of the boot. Raise the rubber matting (if there is any) and check the area immediately behind the rear of the wheel arches. Two large patches of rust will betray a weakening of the rear body mountings.

Fortunately the chassis is a particularly sturdy component and isn't prone to any particular difficulties. The most likely shortcoming is a bent member inherited from a recent or long-forgotten accident. Check if the bonnet fits snugly and if it appears to be ill-fitting then this is probably caused by a mal-aligned chassis. Whilst probing underneath the state of the wiring loom may be the cause for alarm; unfortunately, it runs underneath the body and tends to suffer as a result. If the vehicle in question is bought, then re-routing the loom inside the car should be a matter of priority.

Suspension, steering and brakes

The wishbone/torsion bar independent front suspension is extremely hard-wearing and king pins have been known to endure for 100,000 miles, providing they are regularly lubricated. Shock absorbers are less reliable and replacements are available, while it is possible to have the earlier lever types reconditioned.

Steering is likewise fairly trouble-free and extremely precise, giving two and a half turns lock to lock; in view of this only about an inch of play is acceptable. The rack is situated right out at the front of the car and is subjected to much abuse from objects thrown up off the road. The convoluted rubber gaiters on the rack should, therefore, be checked for evidence of splits or cracks.

Brakes are by Girling and don't suffer from any special peculiarities. Fortunately, the Riley RM Club has a good stock of spares which is a major advantage.

Engine, gearbox and transmission

Both the $1\frac{1}{2}$- and $2\frac{1}{2}$-litre engines are, on the whole, very reliable though the smaller unit does have a deserved reputation for limited bearing life. As already noted the style of both power units dates back to the '30s so perhaps it is not surprising to find that they employ white metal big ends and main bearings, though some later $2\frac{1}{2}$-litre models did use shells. The $1\frac{1}{2}$-litre does, however, suffer from rather narrow oil channels and, though they gave perfectly reliable service when the car was new, over the years sludge can accumulate and restrict the flow. Consequently, insufficient oil reaches the bearing surfaces and wear ensues. In view of this their life-span is estimated to be around 40,000 miles and the Riley RM Club recommend that the filter is changed punctiliously every 4000 miles and the oil every 2000. A replacement filter features on this engine while the $\frac{1}{2}$-litre favours a disposable cartridge type.

Acceptable oil pressure readings for the two power units varies; for the $1\frac{1}{2}$ a reading of 40–50psi at running speeds is just fine, while 20psi is by no means uncommon and quite acceptable for the larger engine. Also, one should be alert for a squeaky or suspect water pump on the RMF variant. The unit was unique to this model and replacement parts are virtually unobtainable.

The RM's gearbox is reliable enough, being essentially of pre-war pattern, though synchromesh is fitted on second, third and top gears. The rear axle should be reliable.

Spares

The Riley RM Club must be commended for the spares service that they offer for these models. Their comprehensive list includes such items as new pistons, gaskets and hoses, while a specially made part is the water pump housing for the $1\frac{1}{2}$– and early $2\frac{1}{2}$–litre cars. Other spares sources include E. T. Lundegaard of Gloucester and W. H. Crossland of Chesterfield, Derbyshire. The well-established Riley Motor Club also caters for these cars.

RM spares and services

W. H. Crossland, Walton Hall, Chesterfield, Derbyshire (tel: Derby 73841).

E. T. Lundegaard, Eastington Trading Estate, Stonehouse, Gloucestershire (tel: Stonehouse 6534).

Riley RM Club, K. C. Clifford, Femdale, Tedburn St Mary, Devon EX6 6DE; Riley Motor Club, Arnold Farrar, The Gables, Hinksey Hill, Oxford.

RM performance and consumption ($1\frac{1}{2}$–litre), *The Autocar*, March 27, 1953. Top speed: 74.25mph. Overall mpg: 24 for 1603 miles.

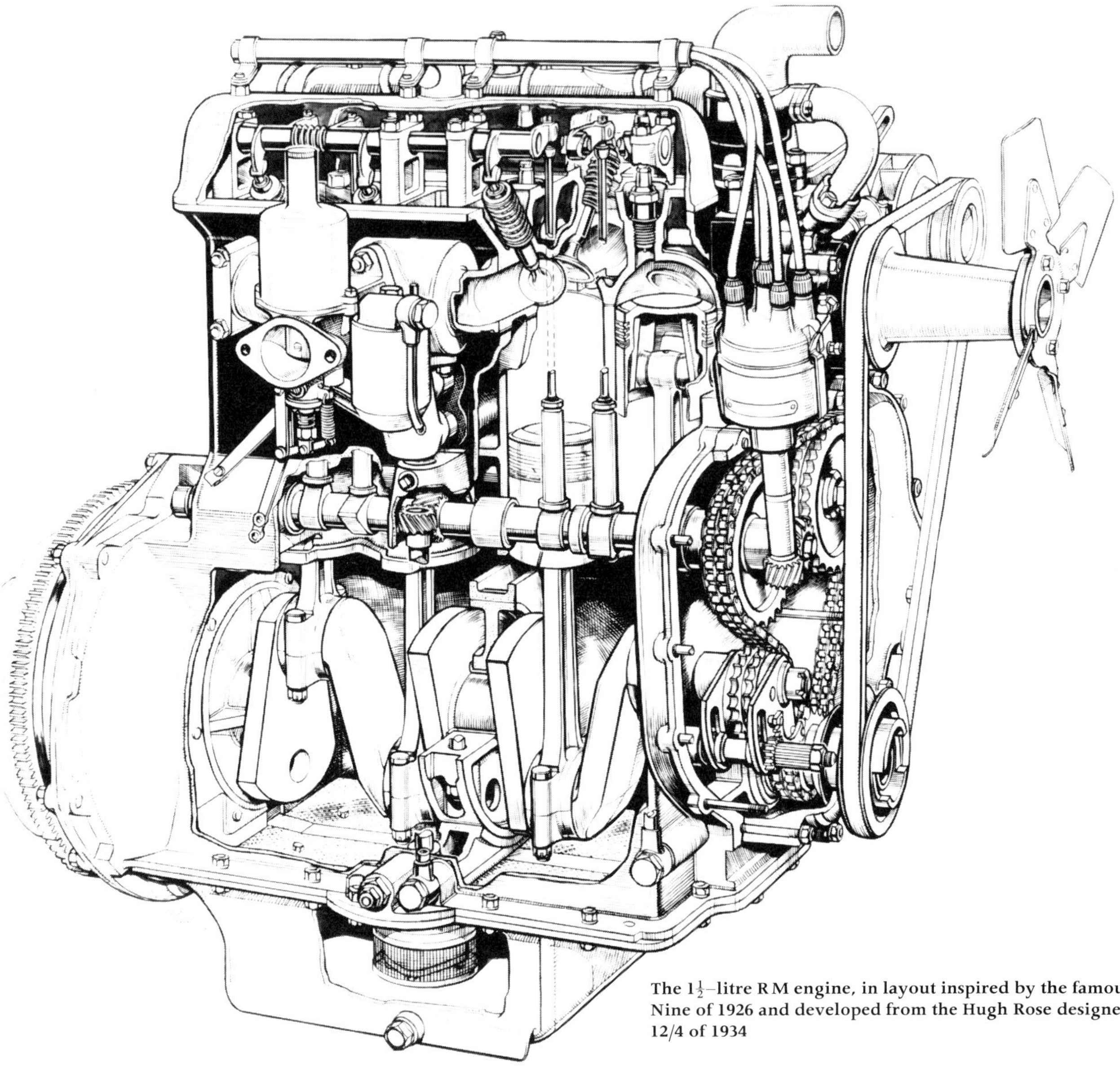

The $1\frac{1}{2}$–litre RM engine, in layout inspired by the famous Nine of 1926 and developed from the Hugh Rose designed 12/4 of 1934

Riley RM $1\frac{1}{2}$ and $2\frac{1}{2}$–litre production

$1\frac{1}{2}$–litre saloon (RMA)		11,854
$1\frac{1}{2}$–litre saloon (RME)		2,096
	TOTAL	13,950

$2\frac{1}{2}$–litre saloon (RMB)		6903
$2\frac{1}{2}$–litre roadster (RMC)		507
$2\frac{1}{2}$–litre drophead coupé (RMD)		500
$2\frac{1}{2}$–litre saloon (RMF)		1050
	TOTAL	8960

Total RM production: 22,910

Coventry production

RMA		6375
RMB		2314
RMC		121
RMD		2 *
	TOTAL	8812

* possibly $1\frac{1}{2}$–litre

Abingdon production

RMA		5479
RMB		4589
RMC		386
RMD		498
RME		2096
RMF		1050
	TOTAL	14,098

An RM with the lid off. A $1\frac{1}{2}$–litre displaying the Citroën inspired torsion bar independent front suspension. The steering rack, positioned immediately behind the radiator, can also be seen

Rolls–Royce 1904 to present date

Frederick Henry Royce's first car took to the road on April 1, 1904. Two more examples of this 10HP 2–cylinder model were built, one being used by his partner Ernest Claremont and the other by Henry Edmunds, who was a director of Royce Ltd. The latter's ownership was fortuitous for it was Edmunds who was responsible for introducing Royce, the Manchester electrical engineer, to the Hon Charles Stewart Rolls, second son of Lord Llangattock. The meeting took place early in May 1904 at the Midland Hotel, Manchester, and the eventual outcome was the creation of Rolls–Royce Ltd. Rolls, a dedicated motorist at the time, had been selling Panhard and Minerva cars with Claude Goodman Johnson in London's Conduit Street and was looking for a luxury car to boost flagging sales. An agreement, dated December 23, 1904 was later drawn up which stated that C. S. Rolls and Co. would sell all the cars that Royce Ltd. could produce and these would be called Rolls–Royces. In addition to the original 10HP car, 15, 20 and 30HP models were offered and in 1905 a 3.5–litre V8 was available but achieved little success. Then, the following year, in March 1906 Rolls–Royce Ltd. was formed. As a result C. S. Rolls and Co. ceased to exist but Royce Ltd. continued as electrical engineers until 1933.

The company's most famous model — the Silver Ghost — first appeared at the 1906 Olympia Show as a 40/50HP, 6–cylinder model, noticeably superior to its contemporaries in its quietness of operation and clearsighted refinement. Two years later, Claude Johnson, who had become Rolls–Royce's managing director, proposed that the 40/50 be the company's sole model and this policy was adopted until 1922. As demand for Rolls–Royces increased, production outstripped Royce's Cooke Street works so a new factory, geared to the production of one model, was opened at Nightingale Road, Derby in July 1908. It was to be the home of Rolls–Royce cars until 1939.

1910 was a traumatic year. In July Rolls was killed in a flying accident at Bournemouth, though he had ceased to be a Rolls–Royce director that April and later in the year Royce was taken seriously ill, the legacy of a life time of irregular eating and overwork. Claude Johnson later took him to convalesce in the South of France and Royce was so impressed with the area that Johnson arranged for the purchase of land at the village of Le Canadel where he also had a house. By 1912 a villa named La Mimosa had been built to Royce's own design and he spent the winter months there until interrupted by the First World War, returning in the autumn of 1921. Summers were spent on the south coast of England at St. Margarets Bay, near Dover and later at West Wittering, Sussex. But after 1910 Royce never again entered the Derby works. However, with the title of Engineer-in-Chief he continued to dominate design from his seclusion, maintaining a drawing office and staff up until his death on April 22, 1933. This left the day-to-day running of the company in Claude Johnson's capable hands.

The outbreak of the First World War had a far-reaching effect on the company's affairs. At the British Government's behest Royce embarked on the design of aero-engines, the first, the 200HP Eagle, appearing in 1915. It was an aspect of Rolls–Royce's business that was destined to far outstrip any profits made by car production so that by the mid-'30s the company was first and foremost a manufacturer of aero-engines.

During the war, the Silver Ghost, appeared in armoured car form, production for civilian use re-starting after the conflict with ouput continuing until 1925. In 1922 it was joined by a smaller 20HP version sporting an overhead-valve engine. This developed into the 20/25 of 1929 and this evolved to the 25/30, which appeared in 1936. The Ghost was replaced in 1925 by the Phantom 1, also with an overhead-valve engine replacing the earlier model's side-valve unit. Production ceased in 1929. In 1926 Claude Johnson, who had guided the company's affairs so effectively, died. He had made an immeasurable contribution to moulding the company's image but much of his energy had been absorbed in an unprofitable venture to build Rolls–Royces at Springfield, Massachusetts, U.S.A. The post of managing director was briefly held by his brother, Basil, but in 1928 Arthur Sidgreaves took over.

The following year the company's new 'flagship', the 7.6–litre Phantom II appeared, though in 1931 it experienced a direct challenge from the 8–litre Bentley, which resulted in a take-over by Rolls–Royce of the bankrupt company. A new generation of RR-based Bentleys appeared from 1933. A new Phantom, the III, appeared in 1936 employing a 7.3–litre V12 engine and was the first

Rolls–Royce to be fitted with independent front suspension. It was an extravagant beast both in manufacturing and maintenance terms and was the last RR model to be built regardless of cost.

Not surprisingly with such excesses, chassis production was unprofitable during the depression of the early '30s and the company were only able to perpetuate Royce's ideology of perfectionist engineering because of increasing profits generated by the aero-engine business. Had Rolls–Royce been solely dependent on chassis production, by the beginning of 1938 the company would have been financially embarrassed, in much the same way as Bentley Motors had been in 1931. Therefore, if the marque was to continue and remain financially viable in its own right, a policy of rationalization was essential.

Early in 1935, Sidgreaves, after returning from a trip to America, felt that the company's products were being overtaken by transatlantic competition and production techniques. So he asked William Robotham, head of the Experimental Department, for proposals for the reduction of costs. Robotham recommended that Rolls–Royce should make much greater use of proprietary parts (at the time electrical equipment and even carburettors were produced by the company) and concentrate on the development of major components. The first outward, but tentative, steps appeared on the Bentley 4¼-litre of 1936 and on the Rolls–Royce Wraith which followed it in 1938. A further aspect of this policy was realized in 1937 with the decision to produce a rationalized range of Rolls–Royce and Bentley cars and the following year the approval was given for the design of a new range of engines. The intention was to produce a power unit in 4–, 6– and 8–cylinder form, sharing a common bore and stroke and therefore pistons and connecting rods. They were designated B40, B60 and B80 respectively. The layout chosen was an overhead inlet/side-exhaust combination which, by chance, echoed the design of the first Royce engine of 1904. Adopting the single overhead inlet valve meant that the thin metal bridge between the valves in the current engine, which had proved a limiting factor to further development, could be eliminated. Although one of the B60 engines was used in the Bentley Corniche of 1939 the full effects of this work would not be felt until after the war, the 6-cylinder unit powering practically all Bentley and Rolls–Royce cars until 1958. (It is probably interesting to recount that the 4– and 8–cylinder versions were widely used by the military, the B40 in such vehicles as the Austin Champ and Dennis Ambulance while the B80 was used to power the Alvis Saracen armoured car.)

But in the late '30s the company was much absorbed in the development of the new Merlin V12 aero-engine and at the instigation of the Ministry of Aircraft Production a new factory was built at Crewe in 1939, thus expanding production.

During the Second World War, under the leadership of Ernest Hives, director and general manager from 1936 to 1945, Rolls–Royce aero-engines were produced by the thousand, Merlins in particular, powering such aircraft as the Hurricane, Spitfire, Mustang, Mosquito and Lancaster. Meanwhile, the car division under Robotham developed the Meteor tank engine, derived from the Merlin, though its manufacture went to Rover in 1943 in exchange for jet-engine development.

When production re-started after the war, Ernest Hives (he became Lord Hives in 1950) was appointed chairman and managing director and car production was centred on Crewe. This left Derby to concentrate on the jet engine that was to

A coachbuilt Silver Dawn drophead coupé, in this instance by Park Ward, and only available for export

revolutionize aircraft propulsion in the postwar years. The company's two postwar vehicles of 1946 were the Bentley Mk VI and the Rolls–Royce Silver Wraith, both employing the $4\frac{1}{4}$–litre's B60 6–cylinder engine initiated before the war.

The Bentley was offered with standardized saloon bodywork and was a pointer to future developments. Back in 1930 the company had contemplated all-steel bodywork and these ideas were sharpened by the far more realistic approach to car manufacture from 1935 onwards. At that time, of course, Rolls–Royce only built their cars in chassis form, the bodies being provided by such coachbuilding establishments as Barker, Hooper and Thrupp and Maberly. But during the '30s Park Ward coachwork graced most Derby products and the close ties between the two companies were consolidated when, in 1938, Rolls–Royce took over the coachbuilding concern. This ensured that individual bodywork could still be manufactured but the accent on postwar products had to be on quantity. And this meant pressed steel bodywork as adopted by the rest of the motor industry from the early 1930s.

Although the Silver Wraith was made available in chassis form only, the Silver Dawn of 1949 was, in effect, a Mark VI Bentley with a Rolls–Royce radiator and was initially only available for export. But it is significant in that it pre-empted the arrival of a common bodyshell for the two marques, which did not happen until the following decade. Both models benefited from a larger 4.5–litre engine in 1951, Wraith and Dawn production ceasing in 1955. In 1950 the Phantom IV appeared powered by a 5.6–litre, B80 straight eight engine, but it was supplied only to royalty and heads of state. It remained available until 1956.

The Silver Cloud (and equivalent Bentley S1) with common bodywork appeared in 1955 and like the Silver Dawn was a Pressed Steel saloon, though right up until the appearance of the Silver Shadow in 1966, chassis were still available for coachbuilt bodywork. The engine represented the final development of the B60, appearing in 4.9–litre form, as pioneered on the Bentley R type Continental of 1954. It was a much bigger car than previous models, indicating a re-designed chassis. Automatic transmission which had been optional on the cars from Crewe since 1952, was standardized.

In 1959 Rolls–Royce purchased H. J. Mulliner, who had been supplying bodies since the days of Charles Rolls, and in 1961 the company's specialist coachbuilding division was re-named Mulliner Park Ward Ltd., though the names finally disappeared in 1971 when the group was re-titled Rolls–Royce Motors Ltd.

But the major event of 1959 was arrival of the Silver Cloud II (Bentley S2) with power steering, very similar to its predecessor, though power came from a new, all-aluminium, 6.2–litre, V8 engine, weighing no more than the 6–cylinder unit it replaced. The Silver Cloud III (Bentley S3) of 1963 had distinctive horizontally-mounted headlights and a more powerful V8. In 1966 came the Silver Shadow (Bentley T series). It marked a significant milestone in the company's evolution with monocoque bodywork, all-independent suspension and disc brakes all round, every feature a first for Crewe. By contrast the V8 Phantom V of 1959–1968 employed only coachbuilt bodywork.

It is interesting to reflect that between 1966 and 1977 more Silver Shadows were built than the *entire* Rolls–Royce output between 1904 and 1939. The foresight of the company's management in the mid-'30s had ensured the survival of a unique automotive institution.

Rolls–Royce production 1946–1967

Silver Wraith (short)	1144
Silver Wraith (long)	639
Silver Dawn	760
Phantom IV	16
Silver Cloud II (short)	2417
Silver Cloud II (long)	299
Phantom V	832
Silver Cloud III (short)	2044
Silver Cloud III (long)	253
Silver Shadow (to 1968)	1483

Silver Cloud and Bentley S1 1955–1959

Specifications and developmental history

The Rolls–Royce Silver Cloud and equivalent Bentley S1 were introduced in April 1955. Standard bodywork was a 4-door saloon by the Pressed Steel Company, though the chassis was also available for individual coachwork. A new box-section chassis with cruciform centre bracing was used, while trailing wishbone and coil-spring independent front suspension was also introduced on the model. Servo assisted brakes were drum all round–Girling hydraulic at the front and mechanical at the rear. Rolls–Royce cam-and-roller steering was fitted. The engine was a 4887cc (95 × 114mm) six with overhead inlet and side exhaust valves with separate inlet and exhaust porting. Automatic transmission was standardized though a manual gearbox was available at special request. The interior was up to the usual Rolls–Royce standards, with upholstery trimmed in leather and walnut dashboard and door cappings. The price was £4796.

Models for 1957 offered power steering as an optional extra on export cars but it soon became available on cars for the home market. At the same time the engine's compression ratio was raised from 6.6:1 to 8:1 and the twin SU carburettors increased from 1.75 to 2 inches diameter. In September 1957 a long wheelbase version of the Silver Cloud was announced, the modification being effected by Park Ward. This facilitated the fitting of a divider between the front and rear passenger compartments, to appeal to potential owners who employed a chauffeur. The car's wheelbase was thus increased from 10ft 3in to 10ft 7in. All models were discontinued in 1959.

Check points

Body and chassis

It is difficult to overstate the importance of checking the condition of the bodywork and it is, therefore, advisable that the car be placed on a ramp to do so. The car's box-section chassis is, on the whole, remarkably sturdy but it is particularly vulnerable where it sweeps over the rear axle at the *top* of the chassis frame. If it has rusted through then the car should be avoided unless it is the work of a special coachbuilder that would respond to some extensive metallic 'surgery'. At the same time the rear body mountings situated in the same area should be inspected and if they are badly rusted then this is a strong disincentive to purchase.

Left, for the Bentley S series owner with everything. Harold Radford offered the Countryman conversion with a special rear platform for picnics and spectator sports

The Rolls–Royce Silver Cloud, as announced in 1955. Automatic transmission was a standard fitment

The exhaust system should be checked at the same time; it is expensive to renew.

The general state of the bodywork is next. It is most unlikely that a completely rust-free example will be encountered and the following remarks apply only to the standard saloon. The bodywork is largely of steel though the doors, bonnet and boot lid are of aluminium. By contrast, a specially coachbuilt body will probably be all aluminium.

Rust is likely to break out along the bottom of the rear wings, the sills and also along the front wheel arches where the bracing strut must be properly secured. The really vulnerable area on the front wings is the section adjoining the front doors and this point should be carefully checked for evidence of body filler. Any small bubbles here indicate that rust has taken hold.

Suspension, steering and brakes

The vast majority of cars were fitted with power-assisted steering. When driving the car one should be particularly suspicious of a steering wheel that feels unbelievably heavy or particularly light. An 'at a glance' method of confirming if power steering is fitted is to look at the number of belts on the fan. Three belts (two for the dynamo and one for the pump) indicate power steering, while those cars without the system have just one belt.

Any wear on the front suspension manifests itself in the king pins and both ends of the top and bottom suspension yokes. Normally they are lubricated by an automatic system, and the reservoir must always be topped up every 5000 miles with EP90 oil. The small bore pipes are often blocked and if, on rocking the car, the suspension creaks this is the most probable cause. At the same time the state of the steering ball joints should be scrutinized. None of these tests should be hurried as re-building of the front end by a specialist is an expensive business.

The brakes are perfectly reliable, but if the vehicle does not respond to braking then the most likely cause is seized wheel cylinders.

Engine, gearbox and transmission

The engine is an extremely reliable and durable unit. It is by no means uncommon for these engines to last for 300,000 miles, though piston and bearing replacement would probably be necessary every 100,000! There are no less then seven mains, incidentally. Oil pressure is an excellent indicator of the engine's overall condition and when the motor is hot the needle should be in the white area of the gauge. If it behaves erratically then the fault could lie in the electric transmitter unit. But should the needle fail to reach the white area then this low pressure could result in a broken conrod.

The engine may sound rather out of sorts and fluffy. If it does then the exhaust valve clearances, which are set at 0.012in., may have closed up or it may be that the ignition has been incorrectly adjusted. This engine is fitted with a Delco Remy distributor containing *two* sets of contact breaker points which provide accurate timing at high speed. Average oil consumption is around 200 miles to the pint.

Generally, the automatic gearbox was very reliable but on the test drive one should watch for evidence of slipping gear bands or a reluctance to change up.

The rear axle is, as one might expect, not prone to problems.

A final word of warning. If the Rolls–Royce version (the Silver Cloud) is contemplated, then it should be confirmed that it is a Silver Cloud and not a re-radiatored S1. The B prefix on the chassis plate will confirm whether the car started off life as a Bentley.

Spares

Fortunately, mechanical, though not body parts, are still readily obtainable for these cars through the Rolls–Royce authorized distributors. A number of companies, including Frank Dale and Stepsons of London and P and A Wood of Great Bardfield, Essex, will undertake restoration work. Clubs are Bentley Drivers and Rolls–Royce Enthusiasts', details of which will be found on page 125.

Silver Cloud and S series spares and services

Frank Dale (Service) Ltd., 101 Farm Lane, London SW6 1QH (tel: 01-385 9724).
P and A Wood, Corniche Garage, Great Bardfield, Braintree, Essex (tel: Great Bardfield 604).

Performance and consumption (S1), *The Autocar*, July 10, 1955. Top speed: 101mph. Overall mpg: 14 for 261 miles.

Rolls–Royce Silver Cloud and Bentley S1 production

Silver Cloud (short)	2238
Silver Cloud (long)	121
S1	3009
S1 Continental	431

Above, the Silver Cloud's driving compartment with polished walnut instrument panel

Below, the 4887cc overhead inlet side exhaust engine of the Silver Cloud

Rover 1904 to present date

The Rover company's origins go right back to 1877 when John Kemp Starley and William Sutton began manufacturing penny-farthing bicycles and tricycles in a small factory in Coventry. One of their special tricycles was called the 'Rover' to indicate that the machine was ideal for roving or wandering around the countryside. The following year came the Rover Safety bicycle, a two–wheeler with almost identically-sized rims with the already popular chain drive to the rear wheel. Here, in effect, was the bicycle as we know it today. The business flourished, Sutton departed and in 1896 the Rover Cycle Company was formed going on to produce motorcycles in 1902, and progressing to motor cars in 1904. A single-cylinder 8HP model was the initial offering, followed by a four–cylinder model in 1905, a year that saw the 'Cycle' dropped from the company's name. A number of twins and fours followed, though at the end of 1911 all other models were dropped in favour of a new Twelve.

After the First World War the company opted for the light-car market with a 998cc, air-cooled, flat twin Eight of 1920–1925. Sales were badly hit by the Austin Seven and an overhead camshaft 14/45 model, designed by Peter Poppe, failed to save the company's finances. A shareholders' committee meeting in 1928 was clearly dissatisfied as no dividends had been paid since 1923 and they appointed Colonel Frank Searle as managing director. Searle was new to the motor industry but was wise enough to appoint the experienced Spencer Bernau Wilks as general manager in 1929.

Wilks, after education at Charterhouse, trained for the legal profession from 1909 and, at the outbreak of the First World War, joined the army, reaching the rank of Captain. He joined the Hillman Company and by the end of the '20s he was the firm's joint managing director along with John Black.

At the time Spencer Wilks joined it, it would not have been unfair to describe Rover as a sinking ship. The company lost £80,000 in 1931 with the deficit spiralling to £279,000 the following year, which is when Rover probably came closest to closure. Fortunately, Wilks struck up an instant rapport with H. Howe Graham, who joined the company as financial director in 1932, and Lloyds Bank extended the period of credit. Rover's manufacturing base was trimmed with output being centred on two factories. Engines and transmissions were produced at the company's Tyseley plant, which they had occupied since 1919, while final assembly was the responsibility of works in Helen Street, Coventry, re-named the New Meteor Works in memory of bicycle days. The 1934 range of cars saw the result of Wilks' careful and informed thinking. Ten and twelve horsepower cars with a common overhead valve, 4–cylinder engine with 100mm stroke were introduced and designated P (for Project) 1. In 1936 a dividend was paid, the largest since 1922, and output and profits increased so that, by 1939, Rover were recording in excess of £200,000. In fact car production more than doubled between 1933 and 1939, and was further re-inforced by the appearance of a new P2 range for 1937. This year also saw the appearance of a 16HP model with a 6–cylinder engine of common parentage to the four.

But international events were to arrest the company's progress. In 1936 Rover were asked by the government to take over a shadow factory for aero-engine production — a measure of the company's stature. Unlike other such factories established throughout the industry, the Rover plant did not adjoin their own premises but was about 20 miles away at Acocks Green, quickly becoming operational in 1937. Soon another shadow factory was needed and this was established on a 65-acre site at Solihull. Again it was built and equipped at government expense and Rover were called upon to organize and manage it. Very prudently, the company decided to purchase a further 200 acres of farm land around the plant, an instance of clearsightedness that was to reap benefits in the expansionist postwar years. From 1940 Rover were involved in the development of the Whittle jet engine, though in 1943 Rolls–Royce took over their work, Rover receiving the manufacturing and development rights of the Rolls–Royce Meteor tank engine as compensation. This power unit, along with the V–8 Meteorite that sprung from it, was manufactured at the company's Acocks Green plant, long after the war.

With the end of hostilities Rover concentrated their activities at the former Solihull shadow factory, having sold their Helen Street premises in Coventry. Tyseley, however, continued as an engines and transmission plant. Production was started up again at the end of 1945, Rover offering their pre-war P2 range of 10, 12, 14 and 16HP models. Gordon Bashford also started work on a new generation of Rovers, though the first project to appear was in complete contrast with later development. Conceived by Spencer Wilks in 1945 the M1 2–door, 4–seater saloon was running by the following year, but it was dropped by 1947. But the diminutive 699cc 4–cylinder engine that powered it is of interest because it appeared with overhead inlet/side exhaust valve layout and sloping cylinder head, echoing a design featured in an experimental pre-war V6 project. Designer Jack Swaine had opted for this configuration so that the layout could be made as compact as possible to fit inside a P1 bonnet!

Meanwhile, work was continuing apace on a new range of cars. However, though the chassis and engine were well advanced, the appearance of the new full-width bodywork was not yet finalized. This was largely the work of Spencer's brother Maurice, who, after spending 2 years with General Motors in America in 1926–1928 joined Hillman and in 1930 moved to Rover as a technical engineer, soon after his brother. Working with his body designer Harry Loker, Wilks eventually produced a shape that met with general approval but was not ready for production until 1949. Therefore a stop-gap model, the P3, was announced in February 1948. Although this resembled the well-established P2 shape, underneath the bodywork things were very different. Engines were available, like the pre-war cars, in 4– and 6–cylinder form. The M1 overhead inlet/side exhaust layout was perpetuated as was the sloping cylinder head, and both engines shared a common 105mm stroke, the 1595cc four being offered as the 60 while the 2103cc six appeared as the 75. For the first time on a production Rover independent front suspension (a Girling patented system employing coil springs and exceptionally wide lower wishbone with a single upper link) was fitted. The chassis was unusual, stopping short of the rear axle as the P1/2 cars had suffered from limited wheel movement with their underslung frames. Therefore the back end of the rear springs was shackled direct to the body.

The P3 gave Solihull some breathing space, the new P4 finally making its public debut at the 1949 Motor Show. This was initially offered in 6–cylinder 75 form and sported the long-awaited full-width bodywork. The 75 and its derivatives were to prove that Spencer Wilks hadn't lost his touch by offering the traditional formula of Rover's good engineering with comfortable and discreet interior, likely to appeal to customers who couldn't run to a Mark VI Bentley and regarded a Jaguar as too flashy. By the time the P4 range ceased in 1964 over 130,000 had been sold.

There were minor changes from some of the components pioneered on the P3. The engine now sported an aluminium cylinder head with twin SU carburettors instead of the iron head with single twin-choke Solex, though the freewheel economy device — a feature of all Rovers since 1933 — was retained. The gearbox employed a column gear change in place of the centrally mounted lever. The front suspension was mildly re-worked, changes having already appeared on the 1949 P3s. By contrast, the chassis was a completely new massive box-section affair with the rear members sweeping over the rear axle as the stressed end of the P3 had proved a weak point. The new bodywork was a combination of steel and Birmabright aluminium alloy, the latter being employed on the doors, bonnet and boot lid to save weight and economize on steel supplies. Partial running boards were revealed when the doors were opened — a sop perhaps to the traditionalists! Over the years the basic formula remained constant, the only exception being the 80 of 1960–1962 which used the then current Land–Rover engine. The P4 and its numerous permutations will be considered separately.

And then, of course, there was the Land–Rover. Maurice Wilks, who farmed an estate in Anglesey had purchased a Willys Jeep for cross-country work just after the war but the fact that he had to buy a foreign vehicle rankled. It was the absence of *any* competition that caused the Land–Rover to be conceived, initially to provide fill in work before the P4 got into its stride. Originally it used the P3 1595cc 60 4–cylinder engine and the Birmabright aluminium bodywork was used purely on grounds of economy. In retrospect the simple functional lines proved ideal and the rust-free aluminium was a master stroke. This tough cross-country vehicle with its fixed four-wheel drive (later engagement was optional) went into production in 1948 and by 1949 it was outstripping car production. The quarter of a million milestone was reached in 1959, the 500,000 figure in 1966 and the million achieved in 1976. This doughty workhorse, which is produced in a fascinating number of variations, was to prove the Rover company's salvation. Graham Robson neatly sums up the situation in *The Rover Story* (Patrick Stephens, 1977): 'Without the Land–Rover the company might not have survived. Because of it Rover rapidly stopped being

small, and became large.'

It will be recalled that during the Second World War, Rover played a formative role in jet design. Therefore with the advent of peace the company, in conjunction with Leyland Motors, investigated the possibilities of applying the gas turbine engine to the private car. The work was undertaken by the Wilks brothers' nephew Spencer King and Frank Bell, who both joined the company from Rolls–Royce at Derby. The outcome in 1950 was JET 1, the world's first gas-turbine powered car. This much modified P4 achieved over 142mph in 1952 with King at the wheel. By contrast, T3, conceived in 1956, was rather more interesting with 4–wheel drive, a de Dion rear axle and rear mounted engine, and the glass-fibre bodywork styled by David Bache. Then there was the T4 of 1961 employing a P6 bodyshell and front-mounted engine. The Rover BRM's successful appearances at Le Mans in 1963 and 1965 gained valuable publicity for the company, but the insurmountable problems of manufacturing costs and fuel consumption eventually resulted in the demise of the project.

The Le Mans exercise had been the brainchild of William Martin-Hurst who had taken over as managing director in 1962. Spencer Wilks retired

Above, **the last of the P4 line, the 110 of 1963. Most have all steel bodies. Wheel trims were shared with the 3–litre.** *Opposite, top*, **1961 3–litre (P5) and,** *below*, **the 2000 (P6) for 1964, both styled by David Bache**

as chairman at the same time, his brother Maurice taking over from him, and when he died suddenly the following year, George Farmer, who had formerly been joint managing director, took his place. But the effect of younger management was already being felt, with Peter Wilks, another nephew, returning to the company in 1954 after having been involved with the Rover based Marauder project with fellow employee George Mackie. Between 1950 and 1952 a mere 15 of these 75 based sports tourers were produced. Initially returning as production manager on the gas turbines side, Wilks soon moved into the engineering department and in 1964 became the company's technical director.

In the meantime the car range had been extended in 1958 with the arrival of the 3–litre P5. Following the establishment of a design department at Solihull the P5 was styled by a youthful David Bache and used an improved version of the P4's engine, disc brakes were soon fitted on the front

464 CNX

145 FLK

wheels. Front suspension was unusual, the independent system employing laminated torsion bars in the Daimler manner. The P5 was destined to remain in production until 1973 benefiting from the company's Buick-based V8 in 1967, being then re-titled P5B (for Buick).

A completely new model, an unusual move for any British car manufacturer at the time, appeared for 1964. The 2000 model (P6) bore no relationship to any previous Rover, having a new all-square, 1978cc single overhead-camshaft engine. The front suspension was unusual, the elongated wishbones pivoting on the bulkhead, a space-saving layout adopted in the hope that a gas turbine engine option might one day be offered. Disc brakes were fitted all round, inboard at the back, while a de Dion rear axle was employed. The highly individual mechanics were more than complemented by equally distinctive styling, again the work of David Bache. Originally conceived with a swept down bonnet and no radiator grille, Bache had to wait until the SD1 of 1976 to realize that objective.

In 1966 came the 100mph, twin-carburettored version, the TC. In true Rover traditions a larger engine had always been projected for the P6, and between 1963 and 1965 experiments were carried out with 5– and 6–cylinder engines, the latter being called P7. However, the engine eventually used was a Buick-based 3.5–litre light alloy V8 and the story of how Rover came to employ it is well worth recounting.

During the winter of 1963–1964 managing director William Martin-Hurst was in America and visited his business friend Carl Keikhaefer of Mercury Marine. Whilst in his experimental workshop at Fond du Lac, Wisconsin, Martin-Hurst saw an alloy V8 which Keikhaefer had just removed from a Buick Skylark as he intended using it in a boat. As it happened the compact V8 had just been dropped by General Motors, ousted by larger engines with thin-wall iron castings. Martin-Hurst was extremely interested in the power unit as he was certain that it would fit the 2000. In the face of opposition from his fellow directors he pushed ahead with the idea and negotiations between Rover and General Motors were initiated in 1964, the licence to manufacture being gained in January 1965. Buick engine designer Joe Turley, who was 18 months away from retirement, was whisked over to Solihull to supervise the anglicization of the engine and the V8 was in production by the autumn of 1967. The principal difference between the Buick and the Rover designs lay in the production of the block. The original version was a gravity die casting but Rover opted for a sandcasting by Birmingham Aluminium, while items like carburettors and electrics were changed as a matter of course. Not that the V8 was initially used in the 2000! The ageing 3–litre was in need of a face lift and it received a mechanical one in 1967, the V8 being fitted in the 2000 in 1968 and titled the P6B. Thus, Rover and, in the longer term, British Leyland were to greatly benefit from Martin-Hurst's chance encounter in Wisconsin.

The '60s were a decade of merger and, though Rover had conducted on/off negotiations with Standard in the '50s, they had come to nothing and Standard were engulfed by an expansive Leyland Motors in 1961. In 1965 Rover were involved in a take-over bid of their own after an approach by Alvis, who by this time were principally manufacturers of aero-engines and military vehicles. However, the same year Rover were shocked when Pressed Steel, who had supplied them with bodywork since the early '30s, was taken over by the British Motor Corporation, who might then be regarded as a competitor. Meanwhile, a new generation of models was taking shape behind the scenes. In 1967 came the P6BS, an exciting mid-engined sports car employing the 3.5–litre V8, designed by Gordon Bashford and Spen King. Then there was a new station wagon that emerged as the highly successful Range Rover of 1970, while ideas were already being mooted for a new P8 saloon intended to replace all the existing Rover models.

But events overtook some of these projects. In July 1966 BMC took Jaguar under its wing to form British Motor (Holdings) Ltd., and Rover were beginning to feel somewhat vulnerable and, at the end of 1966, Leyland Motors approached the company with a view to take over. The offer was accepted and details finalized in March 1967. A new era for 'One of Britain's fine cars' was beginning.

The angular surround and recessed headlamps of the pre-1957 P4s are clearly shown on this 1953 90. *The Autocar* achieved 84mph in this car which cost £1297

Rover production 1946–1967

Twelve		
Saloon		3880
Sports saloon		760
Tourer		200
	TOTAL	4840
Fourteen		
Saloon		1320
Sports saloon		385
	TOTAL	1705
Sixteen		
Saloon		3200
Sports saloon		950
	TOTAL	4150
60 (P3)		
Six light saloon		911
Four light saloon		363
	TOTAL	1274
75 (P3)		
Six light saloon		5196
Four light saloon		2630
	TOTAL	7826
3–litre (P5)		48,548
2000, 2200 and 3500 (P6 and P6B) (to 1976)		327,208

75 1950–1954
60 1954–1959
90 1955–1959
75 1955–1959
105S 1957–1959
105R 1957–1958
100 1960–1962
95 1963–1964
110 1963–1964

Specifications and developmental history

The Rover 75 4–door saloon was introduced at the 1949 Motor Show. A box-section chassis with coil spring and wishbone independent front suspension and radius rods were employed. Girling Hydro-mechanical drum brakes (hydraulic at the front and mechanical at the rear) were fitted. Steering was Burman Douglas re-circulating ball unit. The engine was a 2103cc (65 × 105mm) 6–cylinder with overhead inlet, side exhaust valve layout developing 75bhp. The 75's interior featured wooden glove compartment lids and window surrounds, while the seats were upholstered in unpleated leather. Price was £1006.

Later in 1950, in order to assist air circulation the number of bars in the radiator grille were reduced while the headlight surrounds were painted instead of chromed. The 1951 season witnessed the introduction of full hydraulic brakes while circular instruments replaced the original rectangular ones. In 1952 the radiator grille was altered, the centrally mounted fog lamp, inherited from the P3 was dispensed with and a plain vertically slatted radiator grille replaced the original horizontal bars. At the same time the spare wheel was placed beneath the boot rather than in it. The model was discontinued in 1954. Two new models were introduced for 1954 and for a year overlapped the 75. These were the 4–cylinder 60 with 1997cc (79 × 105mm) engine which developed 60bhp and the 90, a 6–cylinder of 2638cc (73 × 105mm) and developing 90bhp. Improvements over the 75 included a full-flow lubrication system, hardened crankshaft journals and copper lead bearings in place of white metal. Only a single SU carburettor was employed. Although bodily much the same as the earlier model a peculiarity was the hand brake lever, positioned between the driver's seat and the door, as opposed to the shepherd's crook type of the earlier model positioned near the accelerator pedal. The steering column gear change was replaced by a floor-mounted lever, though it was cranked to continue the three abreast front seating. Synchromesh was introduced on second gear.

In 1955 an improved 75 appeared of 2230cc (73 × 89mm) developing 80bhp and was basically a short-stroke 90 unit. On the 75 and 90 front brake drum width was increased to 3 inches. Body changes included raising the boot and rear wing height to increase carrying capacity and a new vertical rearlight cluster with built-in flashers. A three-piece rear window appeared. Inside the car the shepherd's crook handbrake was re-introduced. The following year the 90's compression ratio was raised from 6.73:1 to 7.5:1, producing 93bhp. A Laycock de Normanville overdrive unit was offered as an optional extra, an alternative to the freewheel. The 60 and 75 continued unchanged but all models benefited from new seats and springs with pleated leather upholstery.

Two new models were introduced for 1957, the 105S and 105R. The S was a twin SU-carburettored, high-compression ratio (8.5) version of the 90 engine developing 108bhp while the R used a rather unusual system of semi-automatic transmission. This was made up of a torque converter (with oil being fed from the engine's sump), servo-operated single dry plate clutch, two-speed and reverse manual gearbox and an overdrive that cut in automatically. This layout involved minor chassis 'surgery'. The 60 and 75 continued unchanged but both were now available with an overdrive option instead of the freewheel. All models received a front end face lift. The headlamps, instead of being recessed now projected forward, while flashing indicators replaced the side lights in the tops of the front wings. This meant a new location for the side lights and they were therefore positioned below the flashers. For the 1958 season the range continued virtually unchanged though plain surrounds to the seat centres appeared and a plastic head lining was introduced.

The new P5 3–litre appeared for 1959 and its arrival radically affected the P4 range. A Borg–Warner automatic transmission version duplicated the 105R to some extent so the semi-automatic was dropped, the 105S becoming the 105, there being no need for separate identification. Mechanically, other models for that year remained unchanged, but there were alterations to the body specifications. New bumpers and over-riders of similar pattern to the 3–litre appeared, the radiator grille was recessed, the rear number plate received a die cast shroud and chrome, or stainless steel rubbing strips were fitted. A major change came in 1960. All the previous range was dropped and replaced by two models, the 80, with 2286cc (90 × 89mm) 4–cylinder Land-Rover engine developing 77bhp and the 100 of 2625cc (79 × 92mm). This latter 104bhp six was a short-stroke version of the power unit used in the new 3–litre, having a seven-bearing crankshaft (up until then five had sufficed), roller tappets and a simpler timing chain tensioner. Servo-assisted front disc brakes were introduced on both models and consequently dished wheels with recessed hub caps were fitted to keep the rims within the front wheel arches, and the rear axle was therefore wider. Overdrive was a standard fitment. Both models were largely unchanged at the time they were discontinued in 1962.

Their replacements were the 90 and 110, both powered by 2625cc 6–cylinder engines. The 90 used the 100 power unit but overdrive was not fitted; neither was it available. The 110 employed a new Weslake-designed cylinder head with separate manifolding and twin SU carburettors, as fitted to the 3–litre Mark II. The engine developed 123bhp. Both models were offered with larger instruments, toggle switches and brake fluid and hand brake warning lights, the latter indicators having featured on the later 80 and 100 models. Apart from the very early cars all 95s and 110s have all-steel bodywork in place of the Birmabright aluminium alloy doors, bonnet and boot lid used previously. Production ceased in May 1964.

Check points

Body and chassis

As already mentioned when the Rover 75, the first of the P4 range, was introduced for 1950 it featured steel bodywork but the doors, bonnet and boot lid were aluminium. This continued right up until 1963 when all-steel bodies appeared on the 95 and 110 models, though production was only destined to last another year. Consequently, most of these models are heavier and more likely to rust than

Up until 1954 the range featured unpleated upholstery, as shown on this 90. Note distinctive gearlever and freewheel control on dash

their predecessors. However, it is worth pointing out that *some* examples do retain the aluminium parts. Steel panels were introduced from the following car numbers:

95 Home, right-hand drive	76000493A
95 Export, right-hand drive	76100075A
95 Export, left-hand drive	76300004A
110 Home, right-hand drive	76500972A
110 Export, right-hand drive	76600066A
110 Export, left-hand drive	76800019A

The only other instance where a P4 has had steel panels fitted is if the vehicle has been involved in an accident. From about 1962 these heavier parts were available as replacements.

The front wings are particularly liable to corrosion and on the post-1957 cars this is most noticeable at the top of the wing around the flashing indicators. Another danger point is just below the headlight unit and about half way along the top of the wing should be examined, while the area alongside the front doors, particularly near the bottom is another point where rust can make its presence felt. Moving back down the car, the doors should be opened and the bottom of the central door pillar checked for rust damage. The bottom of the rear wheel arch at its forward end should also be scrutinized for signs of rusting. One place which may well be in an advanced state of corrosion is the rear apron (steel from 1952), while the spare wheel cover situated nearby tends to suffer, as do the hinges, if not regularly oiled. Should the vehicle under inspection be fitted with a sun roof then the immediate areas should be checked for any evidence of the wooden framework rotting underneath the metal.

Fortunately the P4 employs a sturdy box-section chassis, though it is worth noting that the post-1960 examples are rather more prone to corrosion than the earlier cars. It has never been officially admitted, but it seems likely that a lighter gauge steel was used on the frames of these later cars. The chassis is at its most vulnerable directly above the rear axle while the outriggers (there are three each side) will also repay investigation. The one on the front offside is particularly important as the handbrake is mounted on it.

Suspension, steering and brakes

The front suspension units are durable, though the upper suspension bushes often require replacement at 20,000 mile intervals.

Steering is of the re-circulating ball type and remarkably trouble-free. From the later 80 and 100 models the ratio was lowered, no doubt to assist parking.

Brakes are rather more troublesome. The drum braked cars stop well enough, but the 60, 75 and pre-1956 90 cars can suffer from fade under heavy braking and they aren't particularly efficient in reverse. The later 90 and 105 used trailing shoes and are far better but, as they do not have a self-servo effect, the cars are fitted with a vacuum servo unit. A Clayton Dewandre system appeared on the 90 up to 1959 while a Girling layout was employed subsequently. Of the two systems the Girling is the more efficient as it operates on a permanent depression on both sides of the servo piston, the brake opening one side to the atmosphere. The response is consequently quicker than the Dewandre and this is its advantage. The disadvantage is that, after years of use, the unit can develop an excessive 'thirst' for brake fluid and parts are becoming a problem. Spares are more readily available for the Clayton system and it is also easier to service/repair.

The disc-braked (they're fitted to the front wheels) cars are the most effective, but there are a few drawbacks. The discs can corrode and, on early examples, the edge can break off. Caliper pistons can also rust, particularly if the car has been stored for any period of time.

Like many cars, the P4's handbrake does suffer if neglected. The metal gaiter covering the mechanism often rusts to the detriment of the internals. Brake pipes should also be examined, particularly those that run along the top of the chassis.

Engine, gearbox and transmission

With the exception of the 80 model that used the then current Land–Rover unit, all the other cars are powered by overhead inlet, side exhaust F head engines, available in 4– and 6–cylinder variants. The six was by far the most popular, 115,181 being produced, compared with only 15,161 fours.

As already recounted, there are two basic types, the kind fitted to the 60/75 and 105 cars and the other that powered the 95/100/110 range. The later engines are smoother and quieter and will run for at least 100,000 miles without major overhaul. Even when badly worn they still function reasonably well, but the progress is accompanied by an increasing thirst for oil. Not that they were ever economical in this respect, averaging approximately 450–500 miles to the pint when new and 250 miles per pint for a used example. The increase in consumption is often caused by the hardening of the rubber O rings in the inlet valve guides, which are rather short to begin with. Consequently, the lubricant finds its way into the combustion chamber via this relatively uninterrupted route.

With one exception all the cars are fitted with a sturdy 4–speed gearbox, and the only trouble likely to be experienced on the pre 95/110 models is excessive wear on the layshaft front bearing. The exception is the semi-automatic 105R and the system's main fault is usually evidenced by the presence of oil on the clutch plate and occasionally the torque converter can give trouble. It is worth remembering that it was only in production for a couple of years so parts are not going to be in plentiful supply.

The range uses an extremely strong rear axle, dating back to the '30s in design and the author has never heard of one breaking a half shaft. The oil seals are slightly prone to leakage, but this can often be attributed to a build up of pressure in the axle casing, often caused by a blocked breather.

Spares

Body panels for the P4 are virtually unobtainable but engine spares for the post-1959 cars are often obtainable, through the usual retail channels as the power unit endured for many more years in the Land–Rover. Also brake parts are easier to obtain for these later cars because they share components with the P3 3–litre and 3.5–litre cars. There are two Rover clubs, both of which are useful sources for spares. The older of the two, the Rover Sports Register, caters for all models while the younger Rover P4 Drivers' Guild only looks after this specific model.

Rover Sports Register, Adrian Mitchell, 42 Cecil Road, Ilford, Essex; Rover P4 Drivers' Guild, Colin Blowers, 32, Arundel Road, Luton, Beds.

P4 performance and consumption (75), *The Autocar*, March 18, 1955. Top speed: 84.5mph. Overall mpg: 21.9 for 701 miles.

Rover P4 production

60	9261
75	43,677
80	5900
90	35,891
95	3680
100	16,621
105R	3499
105S	7201
110	4612
TOTAL	130,342

Right, under the bonnet of a 90. _Below_, the four cylinder 1977cc 60 engine of 1954. The overhead inlet/side exhaust layout is common to all P4s, apart from the 80

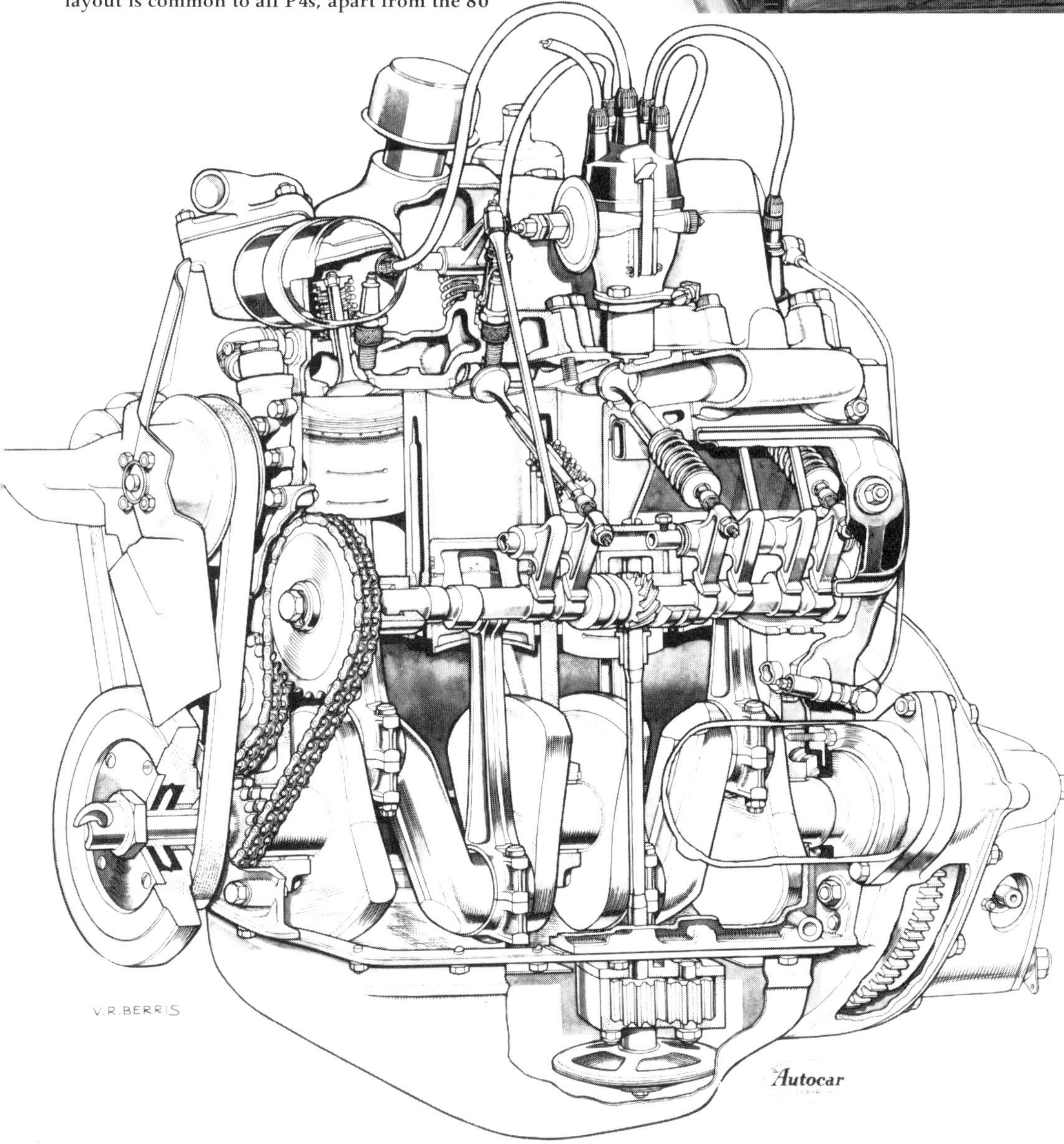

Sunbeam 1899–1937; 1953–1976
Sunbeam–Talbot 1938–1954

The Wolverhampton-based Sunbeam Company entered the car business via the traditional route of the bicycle trade. Their first car, a prototype of 1899, was a 4HP belt-driven vehicle, though the model that reached production was a strange looking device that resembled a Victorian sofa on wheels and known as the Sunbeam Mabley, powered by a single-cylinder de Dion engine and made between 1901 and 1904. Later solidly-built fours were introduced, while 1909 was a significant year as it witnessed the arrival of the Breton engineer Louis Coatalen, who was responsible for the marque's sensational 1,2,3 victory in the French Coupe de l'Auto race of 1912. From thereon the make became increasingly identified with sporting activities.

The year 1920 saw the creation of the unwieldy Sunbeam–Talbot–Darracq group and, while well-built passenger cars continued to be sold, the company's racing programme was stepped up. The marque's most outstanding victory was Henry Segrave's win in the 1923 French Grand Prix, Britain's first GP victory, and he repeated his success at the following year's San Sebastian GP. To generate finance for this expensive programme Coatalen issued a £500,000 8 per cent Guarantee Note in 1924 that was redeemable 10 years later. Although racing was soon downgraded the marque continued to produce their well-engineered and supremely elegant cars throughout the '20s, though the 1.6–litre Dawn of 1934 was hardly in the Coatalen traditions. In 1934 the 10-Year Guarantee Notes were due but the STD group had defaulted on interest payments and there was little prospect of a return on the capital. In October 1934 a receiver was appointed for STD Motors and early in July 1935 Motor Industries Ltd., a subsidiary of Rootes Securities, took over the company. Earlier, in January, they had also absorbed Clement–Talbot Ltd., another member of the STD combine. Badge engineering soon flourished as was the Rootes way of doing things and in 1936 much of the Hillman Aero Minx became the Talbot Ten because the marque had no small car. Sunbeam production came to an end but the Talbot name was continued and for 1938 came the 3–litre which was based on the Humber Super Snipe and that August a new marque, Sunbeam–Talbot was created by merging the two famous names under the Rootes aegis. A 2–litre, based on the 1944cc side-valve Hillman 14 engine was planned for 1940 but after the Second World War, it was re-introduced in August 1945, along with the Ten. Some cars were built at the old Talbot works in London's Barlby Road, but the last car left this factory at the end of May 1946 and from then on production was concentrated at the Rootes factory at Ryton, Warwickshire. The cars were offered mostly in saloon form but there were some

A 1954 Sunbeam Alpine. *The Autocar* managed a top speed of 95mph. It cost £1269 but a heater was £18 extra

touring versions by in-house Thrupp and Maberly and also Whittingham and Mitchel.

At the 1948 Motor Show, the Sunbeam–Talbot 80 and 90 models appeared, though they had been in production since June. The hitherto side-valve engines of 1185cc and 1944cc were provided with overhead-valve conversions, though cart-springing prevailed. Fortunately, Rootes stylist Ted White came up with a superbly elegant sports saloon which perpetuated the pillarless look. A less common coupé version was also offered. Bodies were produced jointly at the company's British Light Steel Pressing at Acton and Thrupp and Maberly. The underpowered 80 was discontinued in 1950 but the 90 was destined for greater things. The Mark II version with larger 2267cc engine received its long-awaited independent front suspension in 1950.

At the instigation of Norman Garrad, who had competed in Talbot 105s in pre-war days, Rootes set up a competitions department and in 1949 won a special prize for the best non-French team in that year's Alpine Rally. The following year things were not so successful, but in 1952 they really hit the jackpot when the team remained intact to win three Coupes des Alpes awards together with the manufacturers' team prize. The year was crowned when the team was awarded the RAC Dewar Trophy for the most outstanding engineering and technical achievement of the year.

The Mark II 90 formed the basis of a convertible modified as a 2–seater by Rootes team driver George Hartwell of Bournemouth. This was achieved by altering the rear of the car and extending the boot. Twin Solex carburettors were introduced and the engine modified to produce 98bhp compared with 70bhp of the original engine. Rootes showed great interest in the project, feeling that the car would have a ready sale on the American market and it was decided to call the model the Sunbeam Alpine in recognition of the 90s successes in the event.

In the meantime a Mark IIA version of the 90 had appeared for the 1953 season. This had larger brakes, perforated wheel discs and improvements to the suspension and steering. The Alpine was based on this model and announced in March 1953, the Rootes prototype having achieved over 120mph at the Jabbeke Autoroute in Belgium, following Jaguar and Rover's wheel tracks at the venue. All the Alpine bodies were produced by Mulliners of Birmingham though the front ends were supplied by British Light Steel Pressings.

The Alpine certainly lived up to its name in 1953 when the team won four Coupes des Alpes and the following year Stirling Moss received a gold cup for achieving a clear run in three successive events. Sheila van Damm's Coupes des Dames rounded off an outstandingly successful crop of Alpine successes. In 1955 came a Monte Carlo Rally win for the 90, a victory that Rootes had just missed in 1952.

The Talbot name was dropped from the 90 saloon for the 1955 season, employing the Humber Hawk version of the standardized Rootes power unit. The Alpine and the the 90 ceased production in 1955 and 1956 respectively. A surprising postscript came in 1957 with the appearance of the Mark IIIS version of the 90. This was produced by Castle Motors of Leicester who had marketed a floor-change conversion for the column-mounted gear change from the Mark I onwards. A louvred bonnet and boot lid hingeing at the top rather than the bottom are external identifying features. About 30–40 were built.

The two seater Alpine in its original 1959 guise. The hardtop was £60 extra, while wire wheels were another £38

In 1956 came the Sunbeam Rapier, based on the Hillman Minx but perpetuating the 90's sporting traditions. Under the bonnet was a 1390cc 'square' engine developing 76bhp. In 1958 came the series II version with larger 1494cc engine, better brakes and floor-mounted gear change, the new version winning that year's RAC Rally. For 1960 came the Series III with an aluminium cylinder head contributing to the 78bhp. The IIIA arrived in 1961 with another increase in engine size to 1592cc and this model lasted until 1963 when it was replaced by the Series IV. This was only available as a two-door saloon. A new five-bearing crank-shaft, 1735cc engine appeared in the Series V model for 1966, the model ceasing production the following year. By contrast the Sunbeam Venezia sports saloon of 1964 based on the Humber Sceptre model, bodied by Touring of Milan, had a brief life, but the restyled Rapier of 1968 lasted until 1976.

A new version of the Alpine, this time a genuine rather than a modified 2-seater, and styled by Ken Howes was announced in 1959. Production was initially assigned to Armstrong Siddeley who in return produced a fairly close copy of their Sapphire engine for use in the Humber Super Snipe. A 1494cc version of the trusted Rootes four was employed, the new Alpine being the first of the company's products to be disc braked. It remained in production until 1968 having reached Series V form.

By this time the Rootes Group had been taken over by the American Chrysler Corporation. The first moves had been made in 1964, Chrysler attaining a controlling interest in 1967. In that year a variation on the Alpine design that had been launched in 1964, ceased production. The Sunbeam Tiger had been the inspiration of Norman Garrad's son Ian, who was Rootes' West Coast sales manager in America. Inspired by the concept of the Ford V8 engined A.C. Cobra, Garrad enlisted the aid of the Carroll Shelby Organization who had been responsible for the Cobra's creation. The outcome was two prototypes, one built by Shelby and the other as a private venture by Ken Miles, who was the organization's chief development engineer. Light 4.2–litre Ford V8s, which benefited from the latest thin-wall casting techniques, were fitted in Series II Alpines and codenamed Thunderbolt. The model was later re-titled Tiger in memory of Sir Henry Segrave's versatile Land Speed Record Sunbeam of 1925. Modification of the Series IV Alpine and fitment of the Ford power unit was entrusted to Jensen Motors and the bulk of cars was exported to America. But the Ford-engined sports car was clearly an embarrassment to Chrysler and anyway the Tiger did not fit into their corporate image, so production ceased in mid-1967.

The Sunbeam Imp Sport, a 51bhp version of the Hillman Imp appeared for 1967 and was supplemented by the Stiletto, a Sunbeam version of the Hillman Californian coupé announced for 1968. But dilution of identity took its toll and Chrysler finally axed the Sunbeam marque in 1976, but revived it as a model name for their new front-wheel-drive cars of 1977.

Sunbeam–Talbot production 1945–1948

Ten	4719
2–litre	1124

Talbot 80 1948–1950
Talbot 90 1948–1954
Alpine 1953–1955
Mark III 1955–1956
Mark IIIS 1957

Specifications and developmental history

The Sunbeam–Talbot 80 and 90 models went into production in June 1948 sharing the same 4–door saloon and drophead coupé bodies and box-section chassis. Suspension was by cart springs front and rear, while drum brakes were Lockheed hydraulic all round. Steering was by Burman Douglas worm and nut. Engines were overhead-valve conversions of the earlier Ten and 2–litre units. The 80, with its nearside manifolding was of 1184cc (63 × 95mm)

The rare and good looking Sunbeam Venezia of 1964, based on the Humber Sceptre. Bodywork was by Touring of Milan

developing 47bhp, while the 90 with its offside manifolding produced 64bhp and was of 1944cc (75 × 110mm). Both were fours. Leather upholstery was used. Price for the 80 saloon was £888 and £991 for the 90.

The 90's front suspension benefited from the removal of three leaf springs from each side and later the steering linkage was altered. The 80 was discontinued in October 1950, a Mark II version of the 90 being marketed for 1951. This employed coil-and-wishbone, independent front suspension and consequently stiffer chassis. Two small grilles either side of the radiator are characteristic of the model and 3 inches were added to the height of the front wings. The larger-bored engine of 2267cc (81 × 110mm), which the model shared with the Humber Hawk, developed 70bhp. A re-designed facia was also introduced. The Mark IIA for the 1953 season had larger brakes, perforated disc wheels with new chrome nave plates and wheel and no rear wheel spats. The Sunbeam Alpine open 2–door, 2–seater was introduced in March 1953 easily identifiable by louvred bonnet, long rear decking and detachable side screens, while the boot lid opened at the top rather than the bottom. The steering was higher geared (2.6 turns instead of 3) and the chassis was re-inforced and stiffened at the front and an additional cross-member introduced. Front suspension was uprated. The 90's engine was mildly tuned to give 80bhp, the compression ratio was raised from 6.4 to 7.4, valve guides were shortened and the inlet manifold was modified.

For 1954 chromium-plated surrounds were introduced on the side grilles adjoining the radiator, while new curved section bumpers were adopted, previously the centre section had served as a number plate mounting. The 90 benefited from the Alpine's higher compression ratio and better steering. The Talbot name was dropped from the 90 for 1955, the model becoming the simpler Sunbeam Mark III. At the front of the car the side grilles were extended to embrace the side lamps, while three small air intakes replaced the 'Sunbeam–Talbot 90' legend on the bonnet side. New slotted wheel discs were fitted to all cars. A slightly more powerful version of the latest Humber Hawk VI engine was used so capacity remained unchanged. Overdrive became an option of the 90 but was a standard fitment on the Alpines sold on the home market. The drophead coupé version of the 90 and the Alpine were dropped in October 1955, but we should not forget the small production Alpine Special with 97bhp engine.

The 90 continued in production in little changed form until it was dropped late in 1956, though a central remote control gear lever had been available as an option to the column change layout from late 1956. Its makers, Castle Motors of Leicester who had offered the modification themselves from Mark I days, produced the Mark IIIS version of the 90 in 1957 after Rootes themselves had ceased production. External features included a bonnet air intake and upward opening boot lid. The engine featured an 8 : 1 compression head with Stromberg D142 carburettor and new inlet manifold. Inside there was a floor gear change, a standard overdrive and Halda average speed indicator reflecting a price of £1350, compared with £1149 for the standard model.

Check points

Body and chassis

All models are prone to body rot, mainly at the rear. Starting there, a special point should be made of examining the rear wheel arches which are double-skinned and particularly vulnerable to corrosion. The spare wheel compartment under the boot also suffers, so the wheel should be removed from its well and the state of the metal checked. Front wings do not rust too badly but investigate around the headlights for signs of trouble. Sills are vulnerable, as are the bottoms of the doors.

The chassis shouldn't present too many problems. It is particularly sturdy and on the Alpine is re-inforced at the front with an extra cross-member, being no less than 14 inches deep at this point. Rear spring hangers should be examined together with the chassis outriggers which incorporate them. Needless to say their role is a vital one.

Suspension, steering and brakes

These components are all substantial units and have no special shortcomings other than those expected with hard-working components. The front suspension is very similar to the Humber Hawk, so spares are easier to obtain, while the Burman worm-and-sector steering isn't prone to any particular problems and is adjustable for wear. Brakes are reliable and spares are still available.

Right, top, **the Sunbeam Mark III of 1955. Enlarged grilles either side of the radiator, wheeltrims and bonnet portholes identify.** *Below,* **Mark III interior. The overdrive control is in the centre of the steering wheel**

PRW 318

The 1184cc Talbot 80 engine of 1949. As it was fitted in the same body and chassis as the 90, the resultant vehicle was rather underpowered

Engine, gearbox and transmission

The engine is a particularly reliable unit and will last for years without any trouble — 100,000 miles without a major overhaul is by no means uncommon. A normal oil pressure reading for these engines is 50psi at 40mph, but lower readings are not necessarily indicative of serious engine deterioration.

However, the gearbox is rather troublesome. It does have a habit of jumping out of third gear and this is often caused by worn detent springs which hold the selectors in place. The box is basically a Hillman Minx unit and is, therefore, under some stress. This, together with lack of maintenance (topping up with oil is difficult), contributes to the problems.

The rear axle is robust, but early Alpines have been known to suffer from cracked differential cages.

Spares

For such a comparatively old range of models, mechanical spares are remarkably plentiful, though inevitably body parts are now virtually impossible to obtain. The main sources of supply are Graham Brooks of London, R. J. Grimes of Coulsdon, Surrey and Suntal Spares of Oldham, Lancs. Owners of these cars can not do better than join the Sunbeam–Talbot Alpine Register which caters exclusively for these cars.

Spares and services

Graham Brooks, 19 William Street, Leyton, London E10 (tel: 01-556 6401).
R. J. Grimes Ltd., Hadleigh Garage, Marlpit Lane, Coulsdon, Surrey (tel: 01-668 1455).
Suntal Cycle and Motor Spares, 47, Rippenden Road, Oldham, Lancashire (tel: 061-624 7409).

Sunbeam–Talbot Alpine Register, John Tolhurst, 201 Havant Road, Hayling Island, Hants.

Performance and consumption (Sunbeam Mark III), *The Motor*, May 25, 1955. Top speed: 93.6mph. Overall mpg: 24 for 559 miles.

Sunbeam–Talbot, Sunbeam Alpine and Sunbeam production

Sunbeam–Talbot 80	3500
Sunbeam–Talbot 90 Mark I	4000
Sunbeam–Talbot Mark II/IIa and Sunbeam Alpine	16,381
Sunbeam Mark III and Sunbeam Alpine	5249

Alpine production is estimated at 2500 to 3000 cars

Alpine 1959-1968

Specifications and developmental history

The Sunbeam Alpine was introduced in July 1959 as a 2–door, 2–seater monocoque with centrally reinforced cruciform. Independent front suspension of the coil-and-wishbone variety was fitted. Brakes were Girling hydraulic with discs at the front and drums at the rear while Burman re-circulating ball steering was employed. A 4–cylinder 1494cc (79 × 76mm), overhead-valve engine, developing 78bhp was fitted. Bucket seats were PVC covered with a shallow bench behind to carry children or luggage. Price was £971.

In October 1960 the Series II Alpine was announced. A larger capacity engine of 1592cc (81 × 76mm), developing 80bhp, was introduced. Repositioned seats and pedals, wider rear springs, and the elimination of eight greasing points were other changes. In March 1961 a Harrington version was announced. This carried a glass-fibre hardtop and, like other Alpines at the time, was available in three stages of tune by George Hartwell. The Series

The Series V Alpine for 1966 with 1725cc engine. The rear fins were cut down in 1964

III version of the Alpine arrived in March 1963. It was offered in two forms, the existing Sports Tourer, boosted to 82bhp with larger inlet valves and the GT with detachable hardtop with less powerful and quieter 75bhp engine. Interior changes to the Sports Tourer included minor alterations to the facia while a new hood was also introduced. Special features of the GT included a wood veneer dashboard, wood-rimmed steering wheel, and a heater was a standard fitment on cars sold on the home market. Both cars benefited from the introduction of servo assisted brakes, the front discs being increased in size and lever type shock absorbers replaced the telescopic ones. Inside there were new front seats with reclining backs and two-speed windscreen wipers and screen washer. A larger capacity boot was achieved by removing the tank from the boot floor and replacing it with two tanks mounted in the wing recesses, giving a total capacity of 11.5 gallons, as opposed to the previous nine. In October the twin Zenith carburettors were replaced by a single twin-choke Solex.

The Series IV Alpine appeared in January 1964. The most obvious external changes were cut-down tail fins and new rear lights. At the front of the car a centrally-mounted badge was introduced in the open radiator intake. Bumpers received rubber-tipped over-riders. The GT lost its less powerful engine, an 82bhp power unit being common to both models. The Borg–Warner model 35 automatic transmission was offered as an optional extra. In September an all-synchromesh gearbox was introduced on both models and a two-speed heater motor was fitted in the GT.

The Series V Alpine was announced in September 1965. It was powered by a 1725cc (81 × 82mm) engine with new five bearing crankshaft and developing 92bhp, and an alternator was standard. Fresh-air footwells were introduced. The Alpine was discontinued in January 1968.

Check points

Bodywork

Although the Alpine favours unit construction bodywork it does have a large cruciform reinforcing member underneath. It isn't particularly prone to corrosion but if so affected the vehicle in question should be avoided. Having checked this the front of the car should be investigated. The main problem here is the engine bay valances which rust particularly badly along the line of the wheel arch. In addition the panel behind the front wheel and the close to toe board suffers similarly. The front wings also tend to rust along their bottom edge and where they join the sills. Doors can also suffer along their lower extremities.

The jacking points should be examined. The back ones are under the rear bumper and their attachment to longitudinal members under the boot floor may be a suspect area. If it is necessary to jack the car up at the back, then the effectiveness of these points will readily be proven. The boot and the spare wheel compartment under it should be examined. The wheel should be removed and the state of the surrounding metal checked. Examine also the rear wheel arches, another area on the Alpine that can suffer at the edges.

Top, **a 1960 Series II cockpit. The glove compartment was supplemented by a useful locker between the seats.** *Above*, **the 1592cc engine developing 70bhp**

Suspension, steering and brakes

The independent front suspension units are shared with the Sunbeam Rapier and Hillman Minx, though the shock absorbers are particular to the Alpine. Fortunately, the suspension wears well probably because it was designed for heavier four seaters. At the rear examine the semi-elliptic spring hangers as they may need attention.

Steering is by re-circulating ball and seems to last well. It is probably worth mentioning that it is easier to inspect the steering after removing the brace between the bulkhead and wheel arch. Brakes are reliable apart from the fact that the self-adjusting rear shoes, fitted prior to the summer of 1967, tend to over-adjust themselves, causing binding. Also the handbrake cable can stick, but lubrication is the usual remedy.

Engine, gearbox and transmission

These rugged engines do have a reputation for leaking oil which usually oozes from the joint between the block and the valve closing plates as well as from the rocker box gasket. Meanwhile, lower down the engine, crankshaft thrust washer wear can present a problem. If depressing the clutch stalls the engine while it is ticking over, then you're in difficulties down below. These power units sound rough and rattly, which is mostly attributable to the aluminium cylinder head and rather basic air cleaner layout.

Gearboxes are durable and, on the whole, reliable. Many cars were fitted with overdrive which should be similarly trouble-free and the same applies to the rear axle.

Spares

A surprising number of mechanical parts for the Sunbeam Alpine are still available through the normal retail outlets. It is also well worth joining the Sunbeam Alpine Owners' Club.

Sunbeam Alpine Owners' Club, Justin Harrington, 32, Kingswood Court, West End Lane, London N.W.6.

Alpine performance and consumption (Series II), *The Autocar*, December 2, 1960. Top speed: 96.7mph. Overall mpg: 20.6 for 1937 miles.

Sunbeam Alpine production

Series I	11,904
Series II	19,956
Series III	5863
Series IV	7936
Series IV (all-synchromesh gearbox)	4470
Series V	19,122
TOTAL	69,251

A small number of CKD Alpines was produced but are not recorded here.

Tiger 1964–1967

Specifications and developmental history

The Sunbeam Tiger was announced in April 1964, initially for export only, being largely based on the Series IV Alpine. Independent front suspension retained the familiar coil-and-wishbone layout, but

A Sunbeam Tiger of 1964. Racing and rally drivers Keith Ballisat and Rosemary Smith with this Alpine Series IV based export earner

with modified spring rates and damper settings, though a new Salisbury rear axle and Panhard Rod were departures. Also new was a rack-and-pinion steering gear by Engineering Productions. The Ford V8 engine was of 4261cc (96 × 73mm), developing 164bhp and a new cross-flow radiator was fitted. Other changes included moving the battery from underneath the occasional rear seat to the right-hand side of the boot, and relegating the spare wheel to the floor. Seats were finished in PVC and a wooden dashboard followed Alpine practice. Price was £1445.

After the first 57 cars had been completed the Borg–Warner T10 manual gearbox was replaced with a Ford unit. A Mark II version of March 1967 was identifiable by an egg-crate radiator grille, full-length side stripes replaced the chrome body trim and silver painted wheels had chrome knave plates. A larger capacity V8 was employed, of 4.7 litres (102 × 73mm), developing 200bhp. A wide ratio gearbox was fitted. Production ceased in June 1967.

Check points

Bodywork

The same rusting problems are shared with the Alpine but the Tiger has a few of its own. Underneath the car the central cross-member should be checked. This is drilled to allow the V8's exhaust pipes to pass through and, perhaps inevitably, this is a potential rust spot. At the rear of the car, the rear spring hangers should be examined, while the Panhard Rod mounting is another point to check.

Suspension, steering and brakes

The special front suspension springs weaken on this model and the Sunbeam Tiger Owners' Club hopes soon to be able to supply replacements. Top suspension bushes wear and these are still available in the normal way but the lower ones are only available from the club.

An item specially produced for the Alpine was the rack-and-pinion steering gear. Unfortunately, the bronze bushes at its extremities wear and replacements just aren't available, so the only solution is to have them specially made. Brake parts are common with the Alpine.

Engine, gearbox and transmission

The important thing about the V8 engine fitted in the Tiger is that the engine oil be changed regularly; every 4000 miles should be sufficient. If this isn't attended to the main bearings can run, so if the car under inspection has a rumble deep down this may well be the cause. Fortunately, new shells can be fitted with the engine *in situ*. An oil filter change isn't that straightforward and this important servicing job tends to be overlooked. The engine will probably respond to a new set of plug leads, as it does tend to overheat and limit cable life. 18mm sparking plugs are fitted and the Sunbeam Tiger OC recommend Autolite BF32s. The single Autolite twin choke carburettor is slightly prone to flooding, incidentally. Gearbox problems are rare and, though the Salisbury rear

Above, the Tiger's 4261cc Ford V8 engine. The rarer Mark II was powered by a 4.7-litre unit. *Below*, the interior differed little from the Alpine. Seats were adjustable for rake

axle may become noisy over the years, it is, nevertheless, very reliable.

Spares

The Tiger's main problem is the same as the Alpine: absence of replacement body panels. Engine and carburettor spares are available from Wolfrace Equipment of London, while electrical parts can be obtained from American Automotive of Wallington, Surrey. Also investigating the spares supply question is the Sunbeam Tiger Owners' Club.

American Automotive, 77 Manor Road, Wallington, Surrey (tel: 01-647 4471).
Wolfrace Equipment, 3 Staples Corner, North Circular Road, London NW2.
Sunbeam Tiger Owners' Club, Dave Woolf, 32 Pinewood Avenue, Hillingdon, Middlesex.

Tiger performance and consumption, *Motor*, May 1, 1965.
Top speed: 116.4mph. Overall mpg: 17.8 for 3455 miles.

Sunbeam Tiger production

Mark I/IA	6495
Mark II	572
TOTAL	7067

Triumph 1923 to present date

Triumph, like many other car manufacturers, started off by making bicycles, progressed to motorcycles and so on to 4–wheelers, though in their case they took a little longer than most. The founder of the company was a German Jewish immigrant, Siegfried Bettmann, who started exporting ready-made bicycles from London but soon decided to manufacture them himself. He therefore formed a partnership with a fellow German Mauritz Johann Schulte in 1887 and a move was made to Coventry, which was the centre of the bicycle industry at the time. Bettmann had sold his bicycles as Triumphs in his London days, so the name was perpetuated and the business prospered, being registered as the New Triumph Cycle Company in 1897. Motorcycle manufacture followed quickly with the appearance of Minerva-engined machines in 1902. Although motor cars didn't appear for the next 20 years or so, Bettmann was for a time, around 1911–1912, chairman of the Coventry-based Standard Motor Company, an ironic circumstance in the light of future events. In 1907 the company moved from their original premises in Much Park Street to Priory Street though the Triumph car of 1923 was built in Clay Lane, in a works that had previously produced the Dawson car of 1919–1921.

The first Triumph car was a 1.4–litre side-valve model titled the 10/20. This was supplemented in 1925 by the 1.9–litre 13/30, having the distinction of being the first British car to be fitted with Lockheed contracting-type hydraulic brakes. But the marque didn't make any real impact until the arrival of the Super Seven in 1927, conceived as a challenge to the well-entrenched baby Austin. About 17,000 examples of the Seven and its derivative the Eight were sold, proving to be the best-selling Triumphs of the inter-war years.

From 1933 the company's entire affairs passed into the hands of Lieut. Col. Claude Holbrook, who had been general manager since 1919 and took over from the ageing Bettmann. Triumph decided to move up market but production figures fell and the company moved deeply into the red. By 1935, for instance, losses totalled £250,000.

The Super Nine of 1932 marked the appearance of the Coventry Climax overhead inlet/side exhaust valve engine produced by Triumph under licence. They were built in 4– and 6–cylinder form until 1937; from thereon Triumph-designed overhead-valve engines prevailed. The first of a new generation of models appeared in 1934, the sporting Gloria range being available in 1100cc and 1232cc 4–cylinder and 1476cc 6–cylinder form. As already recounted, Donald Healey joined the company in 1933 initiating a competitions programme which saw the marque appearing in rallies and trials in an endeavour to engage public attention. Perhaps Triumph's greatest pre-war success came in the 1935 Monte Carlo Rally when Jack Ridley, at the wheel of a Gloria, won the Light Car class (the second year running for the company) *and* achieved an overall second placing. Healey was also responsible for the concept of the Triumph Dolomite of 1934, a 2–seater sports car with an engine closely modelled on that of the contemporary 8C Alfa Romeo, but only three were built.

In 1935 final car assembly was transferred to the former White and Poppe engine factory in Holbrook Lane not far from the thriving SS Cars. The following year the motorcycle side of the business was sold off to raise some much needed finance. The buyer was Ariel's Jack Sangster and with the appearance of Edward Turner's Speed Twin of 1937, Triumph Motorcycle's reputation and profits soared. Meanwhile the Triumph cars soldiered on and, in 1936, came a revival of the Dolomite name, the model being fitted with a distinctive 'fencer's mask radiator grille. Then there was also the 4–cylinder 1767cc Vitesse of the same year that was also offered, like the Dolomite, in 2–litre 6–cylinder form. Finally, in March 1939 came the inexpensive Triumph 12, fashioned on the 1.5–litre Dolomite and costing £285.

The '30s had been difficult years for Triumph. At the beginning of the decade they had considered a merger, first with Rover and then Riley; the latter attempt was briefly but fruitlessly revived in 1938. In 1933 accountant H. Howe Graham had been called in on behalf of Lloyds Bank and the following year he joined the Triumph board to protect Lloyds' interests. But Graham was unable to effect the kind of change that he and Spencer Wilks had achieved when they pulled Rover back from the brink in 1932. Although a modest £35,000 profit was recorded in 1937, in June 1939 Lloyds Bank decided to withdraw when their debts totalled £160,000. Howe Graham's firm of Gibson and Ashford of Birmingham were appointed receivers and in August the two Triumph factories and name were sold to the Sheffield engineering and steel-making group Thomas Ward. The Clay Lane factory was let to Armstrong Whitworth while Donald Healey, who stayed on for a time as general manager, sold the Holbrooks Lane works to H. M. Hobson, who were acting as agents of the Air Ministry. Aero-engine carburettors were produced there throughout the war, though Clay Lane was less lucky, being badly bombed in 1940. Therefore, what was, essentially, little more than the Triumph name and Clay Lane site, were sold in 1944 to Sir John Black, czar of the Standard Motor Company, for £75,000. He promptly sold off the factory, or rather, what was left of it. From thereon Triumph would have a new home at Standard's works at Canley, Coventry.

Standard had been in business since 1903 when William Maudslay began producing cars from 'standard' or interchangeable parts. The early '20s were reasonably profitable but deteriorated from 1927 and 2 years later John Paul Black was appointed general manager. Black, who had been running Hillman, along with Spencer Wilks, revitalized the Standard range and by the end of the decade the company were one of the Big Six motor manufacturers. Not surprisingly, in 1934, he was appointed managing director.

In 1936, along with other car manufacturers, Standard had been invited to participate in the government's shadow factory scheme and a big new works, built at public expense but managed by Black, sprang up adjoining Standard's Canley plant. A further factory for the production of Claudel Hobson carburettors soon joined it. Yet another massive million square foot complex was built a few miles to the west at Banner Lane, Coventry, coming under John Black's control and becoming operational in 1940. It was later to be used for the production of the Ferguson tractor, a successful bid by Black for diversification in the postwar years. In 1940 this testy industrialist became chairman of the Joint Aero Engine Committee in charge of shadow factories and in 1943 he

received a knighthood for his energetic involvement in this vital war work. But for all his success Sir John felt ruffled by the progress of the young SS Cars, so ably managed by William Lyons and to whom Standard supplied engines. He wanted his company to have a sporting edge and give the stylish products from Foleshill a run for their money. And this is where Triumph came in.

The first postwar Triumphs to be built at Canley were powered by 1776cc fours, based on the prewar Standard 14 engine but with overhead valves. The 1800 Town and Country saloon of 1946 featured razor-edge bodywork by Mulliners of Birmingham and was designed by Walter Belgrove, who had styled most of the Triumphs from 1933 onwards. Rather different was the Roadster of the same year which had a suggestion of a sporting flavour. Styled by Frank Callaby, at Black's insistence it was fitted with a dickey seat, probably the last series production car to be so attired as it was an option very much in vogue in the 1920s. In 1948 Standard's new wet-liner, 2088cc 4–cylinder Vanguard engine was fitted to the Roadster but this ceased production the following year. The saloon was accordingly re-named the 2000 when it received the new power unit in 1949, while 1950 models were entitled Renown, and carried the Vanguard's chassis, suspension and steering in addition to its engine. This remained available until 1954.

A scaled-down version, the Mayflower, styled by Mulliners' chief body designer Leslie Moore (with front end by Walter Belgrove) and aimed appropriately enough at the New World went into production in 1949. Under the bonnet, rather disappointingly was a Standard Ten-derived, side-valve, 1247cc engine. Independent front suspension was by wishbone and coil, a layout that was also used on the larger Renown from 1950. Prior to that, postwar Triumphs had used a transverse leaf-spring and wishbone design. However, America didn't respond to the 20th-century Mayflower and production ceased in 1953. Model names sometimes have intriguing origins and in the late '40s

The unconventional approach. The Triumph Roadster employed a dickey seat, probably the last series production car to be so fitted. Note steps on bumper edges to aid access

Sir John, who had served in the RNVR during the First World War, was passing through his naval appreciation period, hence the Triumphs Renown and Mayflower and Standard Vanguard.

In 1950 Walter Belgrove's new Roadster, the rather bulbous TRX made its public debut and while reaction was certainly not hostile, the model left something to be desired, perhaps because it had been styled 3 years previously. Sir John seemed no nearer to his ambition of outpacing the Jaguar and in 1951 he attempted a take-over of the Morgan company. Following the rebuff he approved a cheap no-frills sports car, designed to fill the gap between the MG TD and XK120. The car appeared at the 1952 Motor Show, and has retrospectively become known as the TR1. Walter Belgrove was again responsible for the styling and the engine was a 1991cc version of the Vanguard power unit. Independent front suspension was borrowed from the Mayflower, along with the rear axle, while the chassis had pre-war Standard Flying Nine origins. Obviously there were plenty of problems to be ironed out before the TR2 went into production in 1953 and this was successfully accomplished by ex-BRM man Ken Richardson and Triumph's Harry Webster. In 1954 the model won its first major competitive event — the RAC International Rally — the first of many competition successes, the TR3, with distinctive egg box grille, arrived in 1955 and the following year it received front disc brakes, a notable British first. The 2138cc TR4 with Michelotti designed body appeared in 1961 and in 1965 came the TR4A with independent rear suspension. The Vanguard engine was eventually phased out in 1967, that year's TR5 featuring a fuel-injected, 2498cc, 6–cylinder power unit. From 1959 to 1961 the company's Le Mans entries were powered by a specially developed 2–litre twin-overhead-camshaft engine and in the latter year three glass-fibre prototypes carried off the team prize.

By contrast, the Triumph saloons were in the doldrums during much of the '50s, but then Sir John had always envisaged Triumph as the sporting end of the Standard range. Since the Renown had ceased production in 1954 there had not been a saloon on the home market until the Herald was announced in 1959. It was named after managing director Alick Dick's boat (Black having been de-throned in a 1954 boardroom coup) and remained in production for the next 12 years.

The Triumph TR4 of 1961. Though the Michelotti body with wind up windows was new, underneath was the familiar TR3/3A theme

TRIUMPH

Rather surprisingly it employed a chassis, ostensibly to aid assembly from 'knocked down' form, but helpful because it allowed the body panels to be produced in different locations. The company's body building problems have already been recounted in Part 1. Styling was by Giovanni Michelotti, the first of many for Triumph and later Leyland. A forward-opening bonnet section gave excellent access to the engine and front suspension. These were derived from the Standard 10, the power unit being a 948cc pushrod four, while the suspension was by coil and wishbone. More controversial was the swing axle rear layout but an all-independent suspension car for £702 certainly made good marketing sense. Later in 1961 the capacity was increased to 1147cc and again in 1967 to 1296cc, production eventually ceasing in 1971. The similarly-styled Vitesse of 1962 used a 1596cc, 6–cylinder engine, increased to 2 litres in 1966. The model was discontinued at the same time as the Herald.

The Spitfire, a small sporting car based on the Herald's mechanics and again styled by Michelotti, appeared in 1962 in 1147cc form. A Mark 2 version arrived in 1965 with a more powerful engine while the Mark 3 of 1967 boasted a 1296cc power unit. The Mark 4 of 1970 was face lifted and in 1975 it reached its final 1500cc state, remaining available until 1980. A further development of the Spitfire concept came in 1966 with the fastback GT6. This employed the 2–litre 6–cylinder engine used in the 2000 saloon but suffered from questionable handling in view of its swing axle rear suspension. A Mark 2 version with wishbone irs appeared in 1968 but the model was dropped in 1973 because of the controversial American safety legislation.

A 2000 saloon with all-independent suspension, front MacPherson strut and semi trailing arm and coil spring at the rear, was announced in 1963. The 1998cc engine was inherited from the Vanguard Six of 1961. Capacity was boosted to 2498cc and the model was fuel injected in 1968, being designated the 2.5 PI, but carburettors returned with the restyled Mark 2 of 1969. Output continued for a further eight years. The marque adopted front-wheel drive in 1965 with a well-equipped 1300 saloon. From 1970 it became a 1500 and was discontinued in 1973.

The circumstances that led up to the Leyland Motors take over of Standard–Triumph in 1961 have already been recounted and 3 years after the merger Standard was finally dropped as a marque name. What in 1903 had been a proud boast of engineering innovation 60 years later suggested only attainment of the average. Ironically, Triumph, that had all but ceased to exist in 1939, became the first and most important plume in Leyland's cap.

Triumph production 1945–1967

1800 saloon	4000
2000 saloon	2000
Renown I	6501
Renown Limousine	190
Renown II	2800
Mayflower saloon	34,990
Mayflower drophead coupé	10
10 (rebadged Standard Ten)	9907
Estate Wagon (rebadged Standard Companion)	7351
Pennant (rebadged Standard Pennant)	335
Herald 948 saloon and S	76,860
Herald 948 coupé	15,153
Herald convertible	8262
Herald 1200 saloon	201,142
Herald 1200 coupé	5319
Herald 1200 convertible	43,295
Herald 1200 estate	39,819
Herald 12/50 saloon	53,267
Vitesse 1600 saloon	22,814
Vitesse 1600 convertible	8447
Spitfire Mark 1	45,753
Spitfire Mark 2	37,409
2000 saloon Mark 1	113,157
2000 estate Mark 1	7488
1300	113,008
1300TC	35,342
GT6 Mark 1	15,818
Vitesse 2–litre (Mark 1 and Mark 2 saloon)	12,977
Vitesse 2–litre (Mark 1 and Mark 2 convertible)	6974

Opposite, **the Triumph Roadster of 1948; the raised hood would have isolated the dickey seat passengers!**

Roadster 1946-1949

Specifications and developmental history

The Triumph Roadster (series 18TR) was announced in March 1946. It was an open 2–door 3–seater with a rear dickey seat for two extra passengers. Two 3.5-inch diameter tubular members formed the basis of the chassis which was fitted with independent front suspension of the transverse leaf-spring and upper wishbone variety. Girling Hydrostatic hydraulic drum brakes were employed. Steering was by Marles cam and twin roller. The engine was a 1776cc (73 × 106mm) 4–cylinder overhead-valve unit, developing 65bhp. Seats were leather upholstered. The Roadster went on sale for £889.

At the 1948 Motor Show the model appeared with the 2088cc (85 × 92mm), 4–cylinder, overhead valve Standard Vanguard engine which developed 68bhp. The 3–speed gearbox and rear axle from the same model were also specified and the Roadster was re-designated series 20TR. Production ceased in September 1949.

Check points

Body and chassis

A major advantage for a potential Roadster owner is that almost the entire car is made from Birmabright aluminium alloy. The main exceptions are the wings which are of steel. The snag is that an ash frame was used (and in rare instances oak) and this can obviously rot. One can assess the condition of the timber frame by running a hand in turn behind each sill. The main body of the runners can be felt and this shouldn't disintegrate. If it does then it is probable that the other timber parts out of sight are similarly affected and the vehicle should be avoided unless the considerable expense of repairs is no obstacle.

The steel wings are prone to rusting usually about half way along their length and adjacent to an interior splash plate. The rear ones are double-skinned and, if corroded, can present repair problems. The condition of the number plate box should be checked as it is also made of steel.

As the Roadster employs a robust tubular chassis any problems with rusting are minimal, but the undersides of the tubes are fitted with re-inforcing webs and these can deteriorate. As they are only there to prevent the chassis from flexing don't

The Roadster was fitted with two engines during its production life, a Standard Fourteen based unit, and, from 1948, the Vanguard's wet liner four. Here it's a 1949 example

worry if they're in an advanced state of decay as replacement anyway is cheap and straightforward. The only other difficulty may be with the jacking points right at the back of the car. They can rot badly and become quite unusable.

Suspension, steering and brakes

The independent front suspension is fortunately a straightforward transverse leaf-spring and wishbone layout. The wishbone mounting pins do wear but replacements are available through the Triumph Roadster Club.

Steering is a problem as the box tends to wear excessively. This is usually attributed to the fact that it is fitted with a grease nipple and owners could hardly be blamed if they greased it! This Marles unit should, in fact, run in oil. The grease dries out and wear soon accelerates.

The Girling Hydrostatic hydraulic brakes do suffer from seized trailing shoes. Members of the Roadster Club find that the best way of curing the trouble is to pump the brake pedal. This seems to rectify the self-compensating mechanism while exercising the car on private ground.

Engine, gearbox and transmission

As already recounted the Roadster was first fitted with a Standard Fourteen engine converted to overhead valves, later being replaced with the Vanguard wet-liner four.

Problems on the first engine are centred around the aluminium cylinder head, as the water outlets can corrode and the inlet manifold should be checked for cracks. The four-speed gearbox isn't very troublesome, though the linkage of its column change is likely to wear, the layout being a particularly tortuous one. The Vanguard power unit should give little trouble, though the accompanying 3–speed gearbox is a weakness and the column change can stick between cogs. Acceptable oil pressure readings for both engines are 40psi at 30mph while anything below 25psi should be regarded with suspicion.

Rear axle problems are rare. The original engine used a spiral bevel final drive and the Vanguard a hypoid unit.

Spares

One advantage of the Roadster was that it was fitted with an engine and gearbox that were common to other models. The earlier Standard unit was shared with the contemporary Jaguar 1½ litre, though the latter employed a slightly higher compression ratio and was fitted with an SU carburettor instead of the Triumph's Solex. Parts for this are rather more difficult to locate than the later engine which also saw service in the Standard Vanguard and TR range. However, the Triumph Roadster Club have gone to the extent of having parts made if the need arises and that is one very good reason for joining.

Triumph Roadster Club, R. A. Fitsall, 11 The Park, Carshalton, Surrey (tel: 01-669 3965).

Roadster (1800) performance and consumption, *The Motor*, August 28, 1947.
Top speed: 77.2mph. Overall mpg: 21.2 for 127 miles.

Triumph Roadster production

1800		2500
2000		2000*
	TOTAL	4500

*Of these 666 were left-hand-drive versions.

TR2 1953-1955
TR3 1956-1957
TR3A 1957-1961
TR4 1961-1965

Specifications and developmental history

The prototype TR2, retrospectively titled the TR1, appeared at the 1952 Motor Show. The 2–door, open 2–seater body was mounted on a box-section chassis with coil spring and wishbone independent front suspension. Brakes were Lockheed drum all round and cam and lever steering was employed. The engine was a slightly modified Standard Vanguard unit, being 4–cylinder overhead valve of 1991cc (83 × 92mm), developing 75bhp. Seats were PVC trimmed.

Although the TR2, as we know it, appeared at the Geneva Motor Show in March 1953 production did not begin until August. Improvements included a more robust chassis and a 90bhp engine and overdrive was optional. Cars for 1955 benefited by raising the bottom of the doors and a separate sill being inserted between the front and rear body sections. Rear brakes were increased from 9 to 10 inches diameter, in line with the front ones.

The TR3 was introduced in October 1955. It had a new 'egg box' radiator grille and an option of very occasional rear seats. Engine power was increased to 95bhp and twin 1.75–inch S Us replaced the 1.5–inch units of the TR2. From the autumn of 1956 Girling brakes were introduced, the front discs proving a noteworthy British first while the Mayflower rear axle was replaced by a Phase III Vanguard unit. The TR3A, which was an unofficial designation, was announced in January 1958, though it had been available from the autumn of the previous year on the American market.

This 1956 TR3 road tested by *The Autocar* gave a top speed, in overdrive, of 102mph. The 'egg crate' grille identifies

External changes were a new front panel and full-width radiator grille and slightly recessed headlamps. 'Triumph' appeared on the front of the car for the first time. Lockable door and boot handles were introduced. For 1959 an optional 2138cc (86 × 92mm) engine became available. In 1962 the small production TR3B (yet another unofficial designation) was produced for the American market. This was identical to the TR3A but many were fitted with the TR4 engine and gearbox.

The TR4 was announced in September 1961. It featured a new Michelotti-designed full-width body with wind-up windows, while a useful extra was a Surrey top, allowing the roof panel to be removed but leaving the rear screen in place. The TR4 was mechanically similar to the earlier models but the front and rear track was wider and rack-and-pinion steering was fitted. The 100bhp 2138cc engine was standardized, being coupled with an all-synchromesh gearbox, though the 2–litre TR3A engine was optional. In the summer of 1963 Stromberg CD carburettors replaced the SUs. Production ceased in January 1965.

Check points

Body and chassis

There are two quite distinct body styles to be considered, each with its own set of problems! The same basic design was used on the TR2/3 and 3A while the TR4 employed completely new Michelotti bodywork.

We start with the snags associated with the bodywork of the earlier pre-TR4 cars. At the front of the vehicle the front wings should be carefully inspected. Rust can appear along their bottom edges, while the inner wings can break down and it is not uncommon for holes to appear at the top of the engine compartment, caused by dirt thrown up by the road wheels. Open the doors one by one and then carefully move them up and down as the bases of the mounting posts are prone to rusting. The carpet should be raised and the nearby body mounting bolts carefully examined for signs of corrosion. In fact all the mountings tend to suffer and it is not unknown for the bodywork to be 'secured' to the chassis by gravity alone! Other vulnerable points are the door shut faces while the adjoining areas of bodywork behind and below them are particularly notorious for rust. Examine the inner and outer sills as they are subjected to much abuse from road mud. At the back of the car open the boot and check the condition of the floor, particularly around the rear wheel arches.

On the TR4 the first and most important point to check should be the bases of the door shut posts *inside* the car. The carpeting should be raised and the areas around the seat belt mountings checked, and, if there are signs of rusting, the vehicle should be avoided unless it is being sold very cheaply. Evidence of rusting may also be found along the rear wing deck and if in doubt, then look under the rear wheel arches with a torch. Open the boot and closely inspect the top of the sides for signs of rust. At the front of the car open the bonnet and examine the screw lines where the wings are attached to the engine compartment — it is a favourite rust spot. You may also find corrosion around the bonnet hinge mountings.

The chassis is robust but check where the centre cruciform joins the side members and don't forget

***Top*, the TR3's 1991cc wet liner engine. It developed 5bhp more (95 at 4800rpm) than the TR2. *Bottom*, door handles were still internal. Exterior ones had to wait for the TR3A**

to check the outriggers: another potential rust point.

Suspension, steering and brakes

These are areas where little trouble should be experienced. However, one important front suspension check is the alignment of the lower wishbone fulcrum pins to the chassis. This can easily be done visually. The bonnet is opened and the pins, which should be parallel with the outer edge of the chassis, are located while the welding of the mounting brackets should be intact. If either of the pins is out of alignment correcting it is expensive so be alert for this shortcoming — usually an inheritance from a long-forgotten accident. Also uneven tyre wear often betrays worn trunnions and suspension bushes. The rear springs are extremely strong and the accompanying lever type shock absorbers are hard set so the condition of their mountings must be checked.

The steering gear presents few problems (this also goes for the TR4's rack-and-pinion unit) and the same applies to the brakes, the disc-equipped models having obvious advantages.

Engine, gearbox and transmission

The rugged 4–cylinder wet-liner Vanguard engine will last for years without major attention — mileages in excess of 100,000 being quite common. The engine was fitted with no less than four different cylinder heads during its production life and they are all interchangeable. The only likely trouble is a leaking rear oil seal, though fortunately the lubricant seldom finds its way on to the clutch face. The front one suffers similarly. An acceptable oil pressure reading for this engine is 40psi at 2000rpm, though many seem to run happily at 70psi!

The gearbox fitted to the TR2/3 and 3A is a sturdy and durable unit and has a better reputation for reliability than the all-synchromesh box that succeeded it. This latter TR4 unit is susceptible to layshaft bearing wear and the box does tend to hiss when the clutch is depressed, but this is seldom serious. Overdrive was an optional extra, operating on second, third and top gears.

The rear axle, like many components on the car is strong and reliable and shouldn't present any snags.

Spares

Rather like the Jaguar and MG fraternities, TR owners are very well catered for. The great advantage is that you can not only obtain practically all the mechanical parts you need, but also replica body panels that are indistinguishable from the original. Scarce trim items are also available. Cox and Buckles Spares, that grew as an offshoot of the TR Register's parts service, and based in London and Birmingham is therefore an excellent source of supply. TR Improvements is another London-based outlet with a large supply of parts covering the classic range. The active TR Register is worth joining just for the excellent magazine and parts service it offers!

TR spares and services

Cox and Buckles, Market Road, Richmond, Surrey (tel: 01 :878 7949) and 89 Fairfax Road, West Heath, Birmingham 31 (tel: 021-477 7966).
TR Improvements, 19 Carnarvon Road, South Woodford, London E18 (tel: 01-505 3017).

TR Register, Terry Simpson, Beechcroft, Seymour Road, Northchurch, Berkhampsted, Hertfordshire.

Performance and consumption (TR3 hardtop with 1991cc engine and overdrive), *The Autocar*, December 11, 1957.
Top speed: 102mph. Overall mpg: 24.9 for 561 miles.

TR2/3/3A and B/4 production

TR2	8628
TR3	13,377
TR3A	58,236
TR3B	3331
TR4	40,253

TR4A 1965-1967

Specifications and development history

The TR4A, announced in October 1964 had the same Michelotti-designed bodywork as the preceding TR4, but the radiator grille bars were wider apart and wing-mounted side lamps were introduced. The chassis was new with coil-and-wishbone independent front suspension carried over from the previous model, the semi trailing arm and coil spring independent rear suspension echoed the Triumph 2000 saloon. The engine, of 2138cc (86 × 92mm), developing 104bhp, was similar to the late TR4 version. Interior improvements included a wood veneer dashboard and handbrake positioned between the front seats. An American export version was available with live rear axle. Production ceased in August 1967.

Check points

Body and chassis

The body is virtually identical to the TR4 version dealt with in the previous section. But the TR4A used a new chassis to accommodate the independent rear suspension and it suffers from different and rather more serious problems, though these are mostly confined to the suspension units. Fortunately, it doesn't suffer too much from rusting. The worst areas are the rear outriggers which are open box sections. Rust may also be found between the rear of the two suspension bridges and the aforementioned outriggers. The front of the chassis is usually well protected with engine oil but if the vehicle in question has been involved in an accident then rippling behind the front suspension turrets may be observed, but this is difficult to detect without removing the front wheels.

Suspension, steering and brakes

The TR4A does have serious suspension shortcomings, so checks are particularly important and shouldn't be hurried. The front suspension should be examined first as it can, under certain circumstances, collapse. The parts requiring particular

Right, **the TR refined. The wood veneer dashboard was a feature of the 4A though the layout was inherited from the TR4.**

Far right, **the TR4A's independent rear suspension.** *Below*, **the TR4A. The wing mounted side lights and different radiator grille are the most obvious external differences from the TR4**

The TR4A was the last Triumph sports car to employ the Standard based four, by this time of 2318cc, inherited from the late model TR4s

scrutiny are the lower wishbone mountings and the brackets attached to them, but before proceeding further some background information is necessary. When the TR2 was introduced the front suspension units were mounted on turrets and braced against inward collapse by a detachable cross tube. The lower wishbones pivoted on fulcrum pins held in place by brackets either side of the turrets and were further re-inforced by detachable brackets at their extremities. This was employed on all subsequent models and, when the TR4A appeared, the same coil spring and wishbone layout was retained, but Triumph abandoned the lower mountings and replaced them with easily demountable brackets. On the TR4A there is a single bolt attaching these lower fulcrum brackets to their respective chassis mountings (this was increased to two per bracket on the TR5). If this bolt loosens, the resultant fretting and wear can cause them to be torn free of their mountings if triggered by the imposition of a sudden heavy load, i.e. when the brakes are applied in reverse or the result of a sudden swerve. What happens next is what the TR Register describes as 'a sudden and possibly catastrophic failure' of the front suspension and a severe clunking from that area is about the only warning sign. Therefore, special attention should be paid to the condition and the tightness of the bolts in the mounting brackets; also the welds securing the mountings to the chassis should be examined as they have been known to come adrift at this point.

Now to the rear suspension, with its own set of problems. The bolts securing the forward end of the differential housing to the front suspension bridge straddling the chassis have a habit of tearing free of their mountings. The way to detect if this has occurred is to drive the car forwards and then smartly engage reverse and drive backwards. If an ominous 'clonk' is heard from the back then the housing is probably adrift. A solution to the problem is to re-inforce the bridge with a plate but this is an awkward and expensive exercise with the bodywork in position. A fairly uncommon shortcoming, but one worth recording, is that the extreme ends of the bridge can break off. As they retain the rear coil springs it means that the coils will then be resting on the underside of the body. If the TR4A under inspection looks rather low at the rear or grounds easily then this is probably what has happened. The trailing arm mounting brackets can also tear away from the chassis while the alloy arms can fracture, particularly if the car has hit the kerb.

The steering gear and brakes have already been considered in the previous section, being shared with the TR4.

Engine, gearbox and transmission

Engine and gearbox are inherited from the TR4 and check points have already been listed. Although the TR4A employs independent rear suspension the crown wheel and pinion is the same as that fitted to earlier models and is a strong, trouble-free component.

Spares

The same remarks apply as those already made on page 255 relating to the earlier Triumph models.

TR4A performance and consumption (with overdrive), *Autocar*, May 28, 1965.
Top speed: 109mph. Overall mpg: 25.4 for 1386 miles.

Triumph TR4A production	28,465

Wolseley 1899-1975

Wolseley's formative years and Herbert Austin's involvement with them have already been recounted on page 21. It will be remembered that, up until Austin's departure in 1905, the company only produced horizontally-engined cars, though the smaller vertically-engined Siddeley had been made at the Wolseley works since 1903–4. After Austin's departure, vertical-engined models were introduced and the 1907 range varied from a 10HP twin to a 45HP 6–cylinder. An unusual assignment was the production of caterpillar tractors powered by 12HP air-cooled engines for Captain Scott's ill-fated Antarctic Expedition in 1910. By 1913 Wolseley was Britain's largest vehicle manufacturer with 3000 chassis, though Humber were not far behind with 2500. During the First World War lorries and ambulances were initially produced while Renault and Hispano–Suiza aero-engines were built under licence, the latter, a 200HP V8 overhead-camshaft unit being known as the Wolseley–Hispano Viper.

After the war, in addition to their existing factory at Adderley Park, Birmingham, new premises were purchased at Ward End to cope with the anticipated postwar boom. Wolseley's new cars for 1920 reflected their wartime aero-engine work, the 10HP and 15HP models featuring overhead-camshaft engines. A 3.7–litre side-valve six was also produced. In 1922 a water-cooled flat twin appeared and in 1925 the Ten became the 11/22 while the Fifteen was replaced by the side-valve 16/35. Unfortunately the company's finances were in disarray and in October 1926, Vickers, who had owned Wolseley since 1901, appointed a receiver after liabilities had totalled £2,000,000.

William Morris bought Wolseley in 1927 and set about re-invigorating the firm. His commercial

The Wolseley Ten, a variation on the Morris theme. It had a robust chassis and leaf spring suspension. This is a 1948 example, the last year of production

vehicle business, established in 1924, moved into part of the Adderley Park factory and the remainder was sold off. Car production was then centred on Ward End, with overhead-camshaft engines being developed for the parent Morris company and MG. Morris's influence was soon evident in the Wolseley range, the export version of the 21/60 2.5–litre six using a Pressed Steel body and Lockheed hydraulic brakes.

In 1930 came the 1.3–litre, overhead-camshaft Hornet being followed in 1932 by the 2–litre Viper along with 6– and 8–cylinder variants of the 21/60. In 1932 came the twin carburettored Hornet Special offered in chassis form only. The famous Wolseley illuminated radiator badge appeared in 1933 and the following year a 1000cc Nine was joined by another rash of overhead-camshaft sixes. But, in July 1935, Wolseley, like MG, was sold by William Morris to Morris Motors and some necessary rationalization followed. In came Miles Thomas as Wolseley's managing director and out went the expensive overhead-camshaft engines, the last being produced in 1936. Thereafter Wolseleys were offered as up-market Morrises. In 1939 the ample range embraced a new Ten that used Series M mechanics in a separate chassis, a 1.5–litre 12/48 four while the 14, 16, 18 and 25HP models were sixes. The company suffered heavy losses during the Second World War, the Ward End factory being bombed in 1941. Nevertheless, large quantities of Bren Gun Carriers, mines and gliders were produced.

Most of the pre-war range was resurrected in 1945, with the addition of an Eight. All had overhead valves, the Eight and Ten being up market versions of the respective Morris models. In 1947 came the palatial Twenty-Five powered by a 2.5–litre six though it was only in production for little over a year.

All these models were swept aside and replaced by two new ones which appeared at the 1948 Motor Show. These were versions of the Morris Oxford and Six, being re-named the 4/50 and 6/80 respectively. Both models were monocoques with torsion bar independent front suspension. The 4/50 retained the same bore and stroke as the Oxford, but with an overhead-camshaft layout instead of side valves, while the larger car used an all new 2.2–litre, overhead-camshaft six designed by Morris Engine's Tom Brown. The overhead-camshaft Wolseley was a reality once more! These new models were briefly produced at Ward End, but soon in January 1949 production was transferred to Cowley. In 1953 they were joined by the 4/44 being powered by the 1250cc overhead-valve engine from the MG Y type saloon and had new monocoque bodywork with coil-and-wishbone independent front suspension. The overhead-camshaft engined cars were dropped in 1954, a year that saw the 4/44's body shared with the MG ZA Magnette. In 1956 the Wolseley was re-engined with the Magnette's 1489cc BMC B series four, the resultant model being re-named the 15/50. Production ceased in 1958.

A replacement for the 6/80 appeared for the 1955 season. This was the 6/90, which shared bodywork with the Riley Pathfinder announced the previous year, the new BMC C series 2639cc overhead-valve six being employed. For 1957 the coil-sprung rear axle was replaced by conventional semi-elliptics. That year saw the appearance of the 1500 model. Originally conceived as a replacement for the Morris Minor with a choice of A or B series engines, it was eventually fitted with the latter. Employing the Morris Minor floor pan it shared the model's torsion bar independent front suspension and rack-and-pinion steering layout. Production lasted until 1965.

The new BMC four-door Farina-styled saloon first appeared as the Wolseley 15/60 at the end of 1958. Under the bonnet was the 1489cc four, though the model became the 16/60 in 1962 when the capacity was increased to 1622cc. A 6–cylinder version with 2912cc engine and designated the 6/99 arrived in 1959 and power was increased for 1962, becoming the 6/110. Production lasted until 1968.

Meanwhile the revolutionary BMC front-wheel drive Mini had arrived in 1959 and the Wolseley Hornet variant with plusher interior and bigger boot was announced for the 1962 season. The Mark II version with 998cc engine went on sale early in 1963, receiving Hydrolastic suspension for 1965. The 1967 Mark III had wind-down windows and new door handles but the model was dropped in 1969. An up-market model on the Morris/Austin 1100 theme appeared late in 1965, production running until 1967 when it was replaced by the larger-bored 1300 for 1968. It was discontinued in 1973. Yet another badge-engineered model went on sale in 1967. This was the 18/85, a Wolseley version of the front-wheel drive Austin 1800 with power steering. It remained available until 1972. The marque was finally extinguished in 1975 — like Riley a victim of rationalization.

Wolseley production 1948–1967

4/50	10,938
4/44	29,914
15/50	12,552
15/60	24,639
1500 Mark I and II	68,954
1500 Mark III	31,790
6/90	6300
6/90 Series 2	900
6/90 Series 3	5131
6/99	13,119
6/110 Mark I	11,002
Hornet (to 1970)	28,455

The well instrumented panel of the Wolseley 6/80. The steering column gear change permitted three abreast seating. The heater was an unusual feature of the day

***Overleaf*, smart motoring 1948, an early 6/80. The traditional radiator was in deference to Lord Nuffield's wishes**

6/80* 1949-1954

Specifications and developmental history

The Wolseley 6/80 was announced at the 1948 Motor Show, as a 4-door monocoque saloon with torsion bar independent front suspension and Lockheed hydraulic drum brakes. Steering was a Bishop cam unit. The engine was a 2214cc (73 × 87mm) overhead-camshaft, developing 71bhp The seats were upholstered in leather, the front ones being adjustable for height. Wooden door fillets were fitted, while the dashboard was veneer. Price was £767.

Early in 1949 the car's castor angle was reduced

*Morris Six Series MS 1949–1953 shares similar specifications.

LUNCHEONS
Six Eighty
NWL 445

and in July twin spot-lamps were fitted (chassis number 745). In January 1950 telescopic rear shock absorbers were introduced (1361) and in May twin front telescopics appeared (1850) and later the front of the car was strengthened. September 1950 saw the introduction of separate front side-lights (3167) with 1951 cars having a re-designed facia. Early in that year the rear axle ratio was lowered. A divided bench-type front seat appeared in April 1952 (16709).

The Series Two models appearing in November 1952 had a re-designed cylinder head with inclined instead of vertical valves and improved water circulation. Compression ratio was reduced from 7:1 to 6.6:1. A more powerful dynamo was introduced at the same time. Production ceased in October 1954.

Check points

Bodywork

Areas for scrutiny are familiar: the rear of the front wings alongside the doors, the sills, while in bad cases rust may have invaded the point where the rear wings are attached to the bodywork around the wheel arch. The bottoms of the doors also suffer, the rear more than the front as they contain sound-deadening material which soaks up the water and prevents it from escaping. Check underneath the car where the sills adjoin the floor pan as this can produce some nasty surprises. Open the boot and examine the state of the floor at the extreme rear of the car.

Suspension, steering and brakes

The torsion bar and wishbone independent front suspension is a remarkably trouble-free layout, though the rear semi-elliptics may be rather tired and require re-setting. Early cars used Armstrong lever-type shock absorbers, which were inferior to the Woodhead Monroe telescopics used subsequently. It is rather a surprise to find *twin* Woodheads at the front but they're unlikely to present any major problems, though the mountings can give up the unequal struggle.

Steering is Bishop Cam providing no less than five turns lock-to-lock, which doesn't make the Wolseley the easiest of vehicles to park. The box does tend to leak a little oil and any undue stiffness is probably caused by this.

As with most Nuffield products, Lockheed hydraulic brakes are a major advantage. A hub extractor will be needed to gain access to the shoes and, while wheel cylinders are readily available, master cylinder parts may require more diligent pursuit.

Engine, gearbox and transmission

The 6/80's engine is unusual, but in the established Wolseley traditions, is a single overhead-camshaft six. Drive is by shaft, mounted at the front of the engine and is transmitted via a pair of worm gears which are split and spring-loaded to avoid backlash. At the crankshaft end the shaft drives the oil pump while the top end also operates the distributor.

Camshaft operation is virtually identical to that used on the exquisite twin-overhead-camshaft Alfa Romeos of the '30s. The cam operates directly on to hardened mushroom-shaped tappets which are threaded into the stems of special valves, clearance being set by lock nut. When the valves are re-set, two special spanners are required. The peak of the cam should be pointing upwards at the time and a T-shaped spanner is used to straddle the camshaft and turn the tappet. Another C-shaped key is used to hold the valve spring shroud while this is being done. This would not be mentioned here if it were not for the fact that it might be necessary to do this job as the clearance has to be checked every 3000 miles. And the 6/80 is rather prone to burning out its exhaust valves. An inclination to overheat was probably a contributory cause; unfortunately the model is not fitted with a water temperature gauge. The heating tends to distort the thin end of the valve stem preventing it from seating properly.

The Series 2 cars of 1953 were therefore offered with a re-designed cylinder head. Water jacketing was improved and the engine can be identified by an external water rail. This, along with longer valve stems had the effect of increasing valve life from around 12,000 miles to about double that figure. These later cars also have a two-piece cam cover which means that the distributor timing isn't disturbed when setting up the clearances.

This unconventional power unit has its fair share of unusual features. The starter motor is peculiar to it and the twin SUs are fitted with a small starting carburettor operated from the dashboard. Whilst on the subject of mixture control, it is important that the SUs are kept well tuned otherwise the risk of burning out those exhaust valves is increased: rather too rich than too weak! At the other end of the engine you'll find that the oil filter housing is cast as part of the sump, though fortunately the changing process is straightforward. Never overtighten the four securing bolts incidentally, and make sure that the sludge drain is always kept clear. An acceptable oil pressure reading is 60–80psi at running speeds; tickover can be as high as 40psi though anything down to 20psi is acceptable when the car is stationary.

Unlike the engine, the 6/80's gearbox is reasonably trouble-free and it is encouraging to know that the steering column gear lever is a particularly well-engineered component. Even with the passage of time it shouldn't have developed too much slack.

Spares

Parts, both body and mechanical can be a bit of a problem with this particular model so all the more reason for joining the club that has been formed for them, along with their Morris counterpart.

Wolseley 6/80 and M.O. Club, Roger Tennyson, 31 Azalea Avenue, Somercotes, Derbyshire.

6/80 performance and consumption, *The Autocar*, June 20, 1952.
Top speed: 81.5mph. Overall mpg: 18.1 for 174 miles.

Wolseley 6/80 production	25,117

Left, the 6/80's camshaft drive. Also note the valve adjusters which require special spanners. *Below*, not much to see! The battery seems to fill most of the 6/80's engine compartment though the distributor is readily accessible

Notes on production figures

Throughout this book, with one exception, I have given production figures for the makes and models reviewed. These have been obtained from three sources: the manufacturers, the one-make car clubs and existing publications.

To the club members listed below who went to the trouble of producing totals for me and particularly to Anders Clausager, archivist of B.L. Heritage Ltd., who has provided so many of them, I give my grateful thanks. I should add that the complete set of Austin figures is eagerly awaited.

A.C., Barry Bird, Simon Bathurst-Brown, Brian Gilbart-Smith, A.C. Owners' Club.

Alvis, David Michie, Red Triangle Auto Services.

Armstrong Siddeley, G. A. Horrox, Barry Swainson, Armstrong Siddeley Owners' Club.

Aston Martin, *Autocar* Special Publication, Aston Martin, via the Aston Martin Owners' Club.

Austin A30/A35 chassis numbers, B.L. Heritage Ltd.

Austins by Vanden Plas, *Vanden Plas, Coachbuilders* by Brian Smith.

Austin Healey, *Austin Healey* by Geoffrey Healey, and Graham Robson.

Austin Healey Sprite, B.L. Heritage Ltd.

Bentley, *The Rolls–Royce Motor Car* by Anthony Bird and Ian Hallows.

Bristol, E. J. Wooddin, Bristol Owners' Club.

Daimler, Brian Smith, Daimler and Lanchester Owners' Club and *The Daimler Tradition* by Brian Smith.

Ford, Ford Motor Company.

Jaguar, *Jaguar Drivers Club Year Book 1979/80* by Paul Skilleter.

Jensen, John Pellowe, Jensen Owners' Club and T. W. Beards, formerly of Jensen Motors.

Jowett, Edmund Nankivell, Jupiter Owners' Auto Club and the author.

Healey, *Healeys and Austin Healeys* by Peter Browning and Les Needham.

Lotus, Graham Arnold, Club Lotus.

MG, *MG by McComb* and B.L. Heritage Ltd.

Morgan, Peter Morgan, Morgan Motor Company and *The Four Wheeled Morgan* by Ken Hill.

Morris, B.L. Heritage Ltd.

Riley, B.L. Heritage Ltd.

Rolls-Royce, *The Rolls–Royce Motor Car* by Anthony Bird and Ian Hallows.

Rover, Richard Stenning, the Rover Sports Register and *The Rover Story* by Graham Robson.

Sunbeam–Talbot and Alpine, G. F. Simonds, Sunbeam–Talbot Alpine Register.

Sunbeam Alpine, Sunbeam Alpine Owners' Club.

Sunbeam Tiger, *Tiger* by Mike Taylor.

Triumph, *Triumph Cars* by Richard Langworth and Graham Robson.

Wolseley, B.L. Heritage Ltd.

One make clubs and registers catering for British postwar cars

Allard Owners' Club, Miss P. Hulse, 1, Dalmeny Avenue, Tufnell Park, London N.7 (01–607 3589).

Austin Counties Car Club, Ian Wheater, 82 Northfield Lane, Wickersley, Rotherham, Yorks.

A40 Club, Miss Cynthia Hiller, Flat 12, Jasmin Court, 115 Sydney Road, Sutton, Surrey.

Berkeley Enthusiasts' Club. D. J. Price, 17, Cherry Tree Avenue, London Colney, St Albans, Herts.

Bond Owner's Register, 10, Baber Drive, Feltham, Middlesex.

Bond Equipe Register, Dave McDougall, 13, Wulfstan Drive, Long Itchington, Rugby, Warks.

Buckler Register, Malcolm Buckler, Fairy Oak, Regaby, Isle of Man.

Davrian Register, Alan Rodwell, 47a Ross Road, Wallington, Surrey.

Dellow Register, John Temple, c/o Douglas Temple Studios Ltd., 104b Old Christchurch Road, Bournemouth, Dorset.

Elva Owners' Club, Roger Dunbar, 124, Marine Crescent, Goring-by-Sea, Worthing, West Sussex.

Fairthorpe Sports Car Club (plus **Turner and Rochdale Register**), Barry Gibbs, Rose Cottage, The Hollington, Long Crendon, Aylesbury, Bucks.

[Ford] Consul Capri Owners' Club, Peter Owens, 15, Gyllyndune Gardens, Seven Kings, Ilford, Essex.

[Ford] Mark I Consul, Zephyr, Zodiac Owners' Club, Dave Barry, 25a, Junction Road, Burgess Hill, Sussex.

[Ford] Mark II Consul, Zephyr, Zodiac Owners' Club, Dave Debenham, 26, Burwash Road, Plumstead, London S.E.18

Ford 1600E Club, Bert Mainey, 17, Bowland Avenue, Ashton in Makerfield, Wigan, Lancs.

Ford V8 Pilot Owners' Club, Howard Stenning, 65 Butler House, Burdett Road, London E14.

Ford Classic Owners' Club, John Cantwell, 48 Church Road, Manor Park, London E12.

[Ford] GT40 Owners' Club, Bryan Wingfield, South Gibcracks Farm, East Hanningfield, Chelmsford, Essex.

Gilbern Owners' Club, Richard A. Bonnie, Bucklers Hard, 55 Hempstead Road, King's Langley, Herts.

Ginetta Owners' Club, Stephen Greenwood, 10, Crossways, Colne Engaine, Colchester, Essex.

Gordon-Keeble Owners' Club, Mrs A. Lakas, Gordon-Keeble Car Centre, Westminster Road, Brackley, Northants. (Brackley 702311)

Association of Healey Owners, Neil Mackay, 26, Main Street, Burton Joyce, Nottingham.

HRG Association, Ian Dussek, Wellhampton House, Upper Green Road, Shipbourne, Tonbridge, Kent. (Plaxtol 359).

Post Vintage Humber Car Club, David Edgar, The Warehouse, Serpentine Road, Southsea, Hants.

Historic Lotus Register, Victor Thomas, Badgerswood, School Road, Drayton, Norwich. (Norwich 867464).

[Lotus] Club Elite, D. L. Jinks, Chestnut Cottage, Water Lane, Storrington, Sussex.

[Jowett] Jupiter Owners' Auto Club, Steve Keil, 16 Empress Avenue, Woodford Green, Essex. (01-505 2215).

Lagonda Club, Mrs V. May, 68 Savill Road, Lindfield, Haywards Heath, Sussex. (Haywards Heath 72274).

The Land–Rover Register, Tony Hutchings, Bridge Cottage, 11, Tilmore Road, Petersfield, Hants (Petersfield 4977).

Lea-Francis Owners' Club, W. G. Adams, Amberway, Oxhill, Warwick.

Marcos Club, Colin Feyerbend, 61 Middle Road, Higher Denham, Uxbridge, Middlesex.

Metropolitan Owners' Club, W. E. Dowsing, 4, Burham Road, Knaphill, Surrey (Brookwood 4841).

Mini Cooper Owners' Club, Ray Holman, 9, Walesbeach, Furnace Green, Crawley, West Sussex.

Raleigh Safety Seven and early Reliant Owners' Club, Nicholas Symes, Greenfields, Lyons Road, Slinfold, Horsham, West Sussex. (Slinfold 790262).

Reliant Sabre and Scimitar Owners' Club, Roger Tipler, The Old Bakery, 1, Silver Street, Brixworth, Northants.

Rochdale Register, Derek Bentley, 259, Junction Road, Burgess Hill, Sussex (Burgess Hill 41125)

Association of Singer Car Owners, Barry Paine, 41, Folly Road, Wymondham, Norfolk.

Singer Owners' Club, John Oliver, Dormer

Cottage, Woodham Park Way, Woodham, Weybridge, Surrey. (Byfleet 46359).

Sunbeam Rapier Owners' Club, David Parrott, 185 Milton Road, Cowplain, Hants.

Triumph Mayflower Club, Philip Hall, 75, Morley Road, Staple Hill, Bristol.

Triumph Razoredge Owners Club, Stewart Langton, 25 Mawbys Lane, Appleby Magna, Burton-on-Trent, Staffs

Triumph Sports Six Club, Steve Jarmyne, 69 Stanley Road North, Rainham, Essex.

Triumph Stag Owners' Club, June Armstrong, Hart Racing Services, 73/77 Britannia Road, Fulham, London S.W.6 (01-731 3287).

TVR Club, Douglas R. Manuel, 21 Fishers Road, Staplehurst, Tonbridge, Kent.

Vauxhall Owners' Club, Ron Shier, 41 Oxleys Cottages, Haynes West End, Bedford. (Bedford 741215).

Wolseley Register, Dave Allen, Glenville, Glynde Road, Bexleyheath, Kent.

Vanden Plas Owners' Club, C. Dawe, 10, Playses Green, Hambridge, Langport, Somerset.

Bibliography and photographic credits

Magazines and periodicals

The Autocar (now *Autocar*)
Classic Car
The Motor (now *Motor*)
On Four Wheels
Thoroughbred and Classic Cars

Books

A.C. by Martyn Watkins, Haynes/Foulis, 1976

The Alvis Car by K. R. Day, published by the author, sixth impression, 1974

The American Automobile by John B. Rae, The University of Chicago Press, 1967

Armstrong Whitworth Aircraft by Oliver Tapper, Putnam, 1973

Aston Martin 1914 to 1940 by Inman Hunter, Transport Bookman Publications, 1976

Herbert Austin by Roy Church, Europa Publications, 1979

Austin Healey by Geoffrey Healey, Wilton House Gentry, 1977

Automobile Design: Great Designers and their Work, edited by Ronald Barker and Anthony Harding, David and Charles, 1970

The Big Idea, the Story of Ford in Europe by Dennis Hackett, the Ford Motor Company, 1978

British Cars of the Sixties by Doug Nye, Nelson, 1970

The British Motor Industry, 1896–1939 by Kenneth Richardson, Macmillan, 1977

Bristol Cars and Engines by L. J. K. Setright, Motor Racing Publications, 1974

The Bullnose Morris by Lytton P. Jarman and Robin Barraclough, Macdonald, 1965

By Jupiter! The Life of Sir Roy Fedden by Bill Gunston, The Royal Aeronautical Society, 1978

The Car Makers by Graham Turner, Eyre and Spottiswoode, 1963

Cars of the 1930s by Michael Sedgwick, Batsford, 1970

The Cobra Story by Carroll Shelby as told to John Bentley, Motorbooks International, 1965

The Complete Encyclopedia of Motorcars, edited by G. N. Georgano, Ebury Press, 1968

Daimler, 1896–1946 by St John Nixon, G. T. Foulis

The Daimler Tradition by Brian Smith, Transport Bookman, 1972

Frazer Nash by David Thirlby, Haynes, 1977

Healeys and Austin Healeys by Peter Browning and Les Needham, G. T. Foulis, 1973

More Healey by Geoffrey Healey, Wilton House Gentry, 1978

A History of Technology, Volume V, Oxford University Press, 1979

Jaguar Sports Cars by Paul Skilleter, Haynes/Foulis, 1975

The Jensen Healey Stories by Peter Browning and John Blunsden, Motor Racing Publications, 1974

The Leyland Papers by Graham Turner, Eyre and Spottiswoode, 1971

The Story of Lotus 1961–1971 by Doug Nye, Motor Racing Publications, 1978

MG by McComb by F. Wilson McComb, Osprey Publishing, 1979

Memories and Machines: The pattern of my life by Sir Harry Ricardo, FRS, Constable, 1968

The Mini Story by Laurence Pomeroy, Temple Press, 1964

Mini by Rob Golding, Osprey Publishing, 1979

Morgan by Gregory Houston Bowden, Gentry Books, 1977

Morris Cars, the first thirty-five years by Harry Edwards, the Morris Register, 1978

William Morris, Viscount Nuffield by R. J. Overy, Europa Publications, 1976

The Motor Car 1946–56 by Michael Sedgwick, Batsford, 1979

The Motor Car and Politics by William Plowden, The Bodley Head, 1971

The Motor *Guide to Makes and Models* by David Culshaw, Temple Press, 1959

The Motor Industry by G. Maxcy and A. Silberston, George Allen and Unwin, 1959

The Motor Industry of Great Britain 1970, Society of Motor Manufacturers and Traders

Motor Vehicles, a report on the industry, Political and Economic Planning, 1950

Motoring Entente by Ian Nickols and Kent Karslake, Cassell, 1956

Motoring in the 30s by Graham Robson, Patrick Stephens, 1979

My Years with General Motors by Alfred Sloan Jr., Sidgwick and Jackson, 1965

The Life of Lord Nuffield by P. W. S. Andrews and Elizabeth Brunner, Basil Blackwell, 1955

The Nuffield Story by Robert Jackson, Frederick Muller, 1964

Out on a Wing by Sir Miles Thomas, Michael Joseph, 1964

The Power behind Aston Martin by Geoff Courtney, Oxford Illustrated Press, 1978

A Pride of Bentleys by John Adams and Ray Roberts, New English Library, 1978

The Restoration and Preservation of Vintage and Classic Cars by Jonathan Wood, Haynes, 1977

The Rolls–Royce Motor Car by Anthony Bird and Ian Hallows, Batsford, 1977

Rolls–Royce, The Years of Endeavour by Ian Lloyd, Macmillan, 1978

The Rover Story by Graham Robson, Patrick Stephens, 1977

The Standard Car 1903–1963 by J. R. Davey, published by the author, 1967

Tiger by Mike Taylor, Wilton House Gentry, 1979

Triumph Cars by Richard Langworth and Graham Robson, Motor Racing Publications, 1979

The Triumph TRs by Graham Robson, Motor Racing Publications, 1977

Twenty Years of Crewe Bentleys 1946–1965 by Stanley Sedgwick, Bentley Drivers Club, 1973

Vanden Plas, Coachbuilders by Brian Smith, Dalton Watson, 1979

Veteran Motor Car Pocket Book by Anthony Bird, Batsford, 1963

Wolseley by St John Nixon, G. T. Foulis, 1949

Works Team, the Rootes Competition Department by Michael Frostick, Cassell, 1964

Photographic credits

Grateful acknowledgment is given to the following for permission to reproduce photographs and drawings:
Autocar (IPC Transport Press): pages 13, 23, 48, 57, 59 (lower), 60, 62 (upper), 65 (upper and lower), 67 (upper), 71, 78, 80, 82, 87, 90, 97, 99, 103, 104, 108, 110, 111, 113, 115, 117, 121 (middle), 122, 125, 127 (lower), 130 (upper), 132, 133, 137, 142, 145 (lower), 147, 148, 152 (upper and middle), 154, 157, 158, 161, 163, 169, 170, 172, 175, 177, 178, 179, 180, 186, 187, 194, 196, 197, 198, 199 (lower), 214, 215, 223, 229, 231, 233, 235, 239, 240, 242, 244 (lower), 246, 251, 252, 253, 254, 257, 258, 259, 265 (lower).
B. L. Heritage Ltd: 10, 15, 29, 31, 52, 61, 62 (lower), 65 (middle), 65 (upper), 76 (lower), 118 (left), 121 (lower), 183, 185, 188, 191, 193, 202, 206, 207 (upper), 209, 210, 211, 213, 216, 261, 262, 265 (upper).
Ford Motor Company Ltd: 14, 17, 26 (upper), 27, 28 (upper), 55, 139, 140, 141, 143, 145 (upper).
General Motors Corporation: 33
IPC Specialist and Professional Press: 32.
Jensen Owners Club: 63 (top), 165, 166, 168.
Rolls–Royce Motors Ltd: 123, 221.

Index to Part 1

Part 2 models analysed